The Fight Against Doubt

The Fight Against Doubt

HOW TO BRIDGE THE GAP BETWEEN SCIENTISTS
AND THE PUBLIC

Inmaculada de Melo-Martín and Kristen Intemann

OXFORD
UNIVERSITY PRESS

Oxford University Press is a department of the University of Oxford. It furthers the University's objective of excellence in research, scholarship, and education by publishing worldwide. Oxford is a registered trade mark of Oxford University Press in the UK and certain other countries.

Published in the United States of America by Oxford University Press
198 Madison Avenue, New York, NY 10016, United States of America.

© Oxford University Press 2018

Library of Congress Cataloging-in-Publication Data
Names: Melo-Martín, Inmaculada de, author.
Title: The fight against doubt : how to bridge the gap between scientists and the public / Inmaculada de Melo-Martín and Kristen Intemann.
Description: New York : Oxford University Press, 2018. |
Includes bibliographical references and index.
Identifiers: LCCN 2017040303 (print) | LCCN 2018020332 (ebook) |
ISBN 9780190869236 (online course) | ISBN 9780190869243 (updf) |
ISBN 9780190869250 (epub) | ISBN 9780190869229 (cloth : alk. paper)
Subjects: LCSH: Science—Social aspects.
Classification: LCC Q175.5 (ebook) | LCC Q175.5 .M467 2018 (print) | DDC 306.4/5—dc23
LC record available at https://lccn.loc.gov/2017040303

For the children in our families, Martín, Cristina, Christian, and Nicolás, hoping that our work will contribute to a better world for them, where dissent does not obfuscate.

CONTENTS

ACKNOWLEDGMENTS

Like all books, this one has benefited from the generosity and expertise of friends and colleagues. Thanks are due to John Beatty, Evelyn Brister, James Brown, Stephen Brown, Martin Carrier, Sharon Crasnow, Kevin Elliot, Daniel Flory, Karen Frost-Arnold, Maya Goldenberg, Daniel Hicks, Stephen John, Ian James Kidd, Boaz Miller, Kristin Shrader-Frechette, Torsten Wilholt, and Alison Wylie for taking the time to read and comment on earlier versions of the manuscript. We are also grateful to the organizers and participants of the Workshop on the Epistemic Role of Manufactured Dissent in Climate Science at Karlsruhe Institute of Technology, and particularly to Justin Biddle and Anna Leuschener for pushing us to think more carefully about epistemically detrimental dissent. Thanks also to the reviewers of the manuscript for their thoughtful and detailed feedback, and to Lucy Randall, our editor at OUP, for her encouragement and support through all the stages of this project.

This book would not have been possible without a collaboration that now spans years. We have learned to understand quite well how one another thinks and often can anticipate each other's criticisms. We believe that this collaboration has resulted in a better manuscript than it would have been otherwise, and it undoubtedly made the work much more fun. This collaborative work was greatly facilitated by both of our institutions, Weill Cornell Medicine and Montana State University, who gave us time and support to work together—both virtually and in person. Kristen Intemann is particularly grateful to Dean Nicol Rae and the College of Letters and Sciences for travel support. Thanks also to our colleagues in the Division of Medical Ethics and the Department of History and Philosophy for their support and their willingness to discuss many of the ideas addressed in this book.

Last, but certainly not least, we are immensely grateful to and for our family and friends. They provide the love, support, laughs, and encouragement that make life joyful and work meaningful. We owe special thanks to our parents Gerald and Christina Intemann, Elinor and Ken Wymore, Alfredo de Melo Coello, and Catalina Martín Cano. They inculcated in us a love for learning and helped cultivate a sense of social justice. Kristen thanks Pablo Fernandez Coletes whose constant support has brought joy and balance to her life and whose boundless optimism makes her believe that the world can be a better place. Inmaculada is grateful to Martín and Cristina, for bringing so much delight to her life.

The Fight Against Doubt

1

Dissent and Its Discontents

During his time in office from 1999 to 2008, South African president Thabo Mbeki and his Minister of Health, Manto Tshabalala-Msimang, firmly opposed government support for existent HIV antiretroviral (ARV) drugs and sought to limit their use in the country.[1] Despite the overwhelming global scientific consensus that HIV causes AIDS and that ARVs are the most effective way to control the disease, Mbeki's government questioned these claims and withdrew support from clinics that had started using ARVs, obstructed Global Fund grants, and restricted use of donated ARVs.[2] Researchers have estimated that delays in the provision of ARVs in the public sector during the Mbeki presidency resulted in the premature death of at least 330,000 South Africans. Similarly, they have argued that because of restrictions on the use of drugs to prevent mother-to-child transmission of HIV, 35,000 babies were born infected with the virus.[3]

Currently, in several countries including the United States and the United Kingdom, an increasing number of parents are refusing to vaccinate their children, despite the clear consensus among researchers that pediatric vaccines are safe and effective.[4] Parental refusal of the measles, mumps, and rubella (MMR) vaccine during the last two decades has had a devastating effect on public health, resulting in outbreaks in places where the disease had been eradicated. In the early 2000s, vaccination rates dropped precipitously in England going from 91% in 1998 to below 80% in 2003.[5] New cases of measles increased from 56 in 1998 to 1370 in 2008. Measles outbreaks began to occur in London, where less than 50% of children were immunized, and quickly spread to Scotland and Ireland. Similarly, the United States, where measles was thought to have been eradicated since the year 2000, has suffered a resurgence in the disease. In 2014, there were 644 cases of measles, three times as many as reported the year before.[6] Researchers have estimated that MMR vaccination rates among the exposed population in which secondary cases occurred might be as low as 50% and likely no higher than 86%.[7]

There is a wide gap, particularly in the United States, between the results of climate science and public support for climate policies. According to a recent report by the Lancet Commission on Climate Change and Health, the effects of climate change represent a potentially catastrophic risk to human health.[8] Possible direct and indirect health impacts include higher rates of respiratory and heat-related illness, spread of disease vectors, increased food insecurity and malnutrition, and displacement. The World Health Organization estimates that an additional 250,000 deaths per year from 2030 to 2050 will result from climate change related effects.[9] Such changes will also produce increased drought-related water and food shortages, damages from river and coastal floods, and extreme heat events and wildfires. Research shows that all nations will face the adverse environmental, economic, and health effects of climate change, but developing countries and vulnerable populations such as the elderly, children, and persons with chronic illnesses will be disproportionately impacted. Available scientific evidence shows that preventing or limiting these negative effects requires global and immediate behavioral and policy changes intended to reduce greenhouse gas emissions.[10] Nonetheless, although concerns about reducing global warming are part of the international political agenda, it has been difficult to obtain a binding international agreement to curb climate change by all major emitters worldwide. The Paris Agreement, an agreement within the United Nations Framework Convention on Climate Change that aims to address greenhouse gases emissions mitigation, adaptation, and finance, is due to take effect in the year 2020. Although the Obama administration signed and ratified the Paris Agreement, the Trump administration subsequently decided to withdraw from it. While many expressed concerns about this move, many Americans oppose policies directed at tackling climate change and support Trump's decision. The most effective policy implementations have been made by states and cities.

Why do people reject scientific claims that scientific experts accept? What accounts for this gap between what the scientific community takes as broad and well-grounded consensus and people's support for such consensus? And why do people push back against scientifically grounded public policies and actions? A variety of complex factors, including educational level, religious beliefs, political affiliation, and social identity, contribute to the existence of these general differences between the public and the scientific community.[11] However, many have called attention to the role that a handful of dissenters have played in generating these various disastrous consequences.[12]

In justifying his AIDS and ARV policies, Mbeki explicitly seized on the dissenting research of Peter Duesberg and his colleagues who insistently denied that HIV was the cause of AIDS and that ARV drugs were useful.[13] These scientists also served on Mbeki's Presidential AIDS Advisory Panel in the early 2000s to determine whether HIV caused AIDS. In fact, some researchers have directly implicated Duesberg and colleagues—and not just Mbeki's government—in the resulting South African AIDS deaths.[14]

Parents who refuse to vaccinate their children often cite concerns about vaccine safety raised by a discredited study linking the MMR vaccine and autism. In 1998, Andrew Wakefield described a new autism phenotype he argued was triggered by environmental factors such as the MMR vaccine.[15] The publication received a significant amount of attention and, with help from the media and highly visible, sometimes celebrity-led advocacy groups, convinced many parents that they should refuse to vaccinate their children.[16] Although the data presented in the paper have been declared fraudulent[17] and Wakefield has been found guilty of ethical, medical, and scientific misconduct in the publication of the research,[18] the influence of Wakefield's dissent in parent's vaccination decisions has been difficult to counter.

Similarly, climate change skeptics have pointed to instances of dissenting views about anthropogenic climate change as evidence that there is no scientific consensus and that potential mitigation policies are unnecessary or unwarranted. In many cases, private industries have funded such dissenting research and think tanks have spent millions of dollars in publicizing these views that benefit their particular economic and political interests.[19] The media's practice of "balanced reporting"—a perceived duty on the media's part to give equal coverage to dissent and the scientific consensus–has also contributed to the perception that dissenting voices are widespread. Although this practice is increasingly questioned,[20] this so-called balancing requirement has made dissenting views appear more legitimate and extensive than they actually are.[21] Many scholars have therefore concluded that the publicity given to problematic expressions of dissent over the existence of anthropogenic climate change has led to public confusion and doubt, as well as apathy and lack of support for mitigation policies.[22]

As these examples show, some instances of dissent about scientific claims have considerable negative epistemic and social impacts. Dissent can result in confusion or false beliefs about what the evidence says, the existence of a scientific consensus, or the degree to which scientific claims are supported by the evidence. Furthermore, insofar as such dissent pushes scientists to repeatedly critique dissenting voices or justify why such views ought to be dismissed, it can force scientific communities to waste valuable time and other resources. Dissent, particularly when it is aggressive and politically charged, can also create a context in which scientists feel intimidated, slowing or impeding scientific progress.[23] Scientists may avoid research in particular areas or water down their conclusions in order to avoid reprisals. Some cases of dissent, then, not only can fail to contribute to the advancement of science but actually can hinder it.

When scientific evidence is relevant for policy or action, these negative epistemic consequences can also have other adverse social impacts. Scientific consensus is a benchmark for scientific knowledge[24] and thus essential for legitimately grounding policy decisions. Scientific and technological expertise is important for legitimizing public policies, health advice, medical treatments, or regulatory decisions.[25] When experts agree on a particular scientific claim—for example, that smoking causes cancer, that anthropogenic causes are responsible for global warming, that condoms prevent transmission of HIV, or that childhood vaccines are safe—such consensus lends validity to the public policies supported by such scientific knowledge, such as restrictions on smoking, CO_2 regulations, free provision of condoms to those at higher risk of contracting HIV, and mandatory vaccination. To the extent that dissent appears to challenge the scientific consensus, it can lead the public and policymakers to doubt the strength of the evidence in support of particular public policies or behaviors, and cause them to reject and actively oppose such polices.[26]

How should we go about limiting the damage that problematic dissent regarding scientific claims can have? The purpose of this book is to answer this question.

One way to address this concern would be for scholars to point out features of dissent that make it problematic. Several authors have identified common strategies that some dissenters have employed to manufacture doubt,[27] that create the appearance of scientific controversy,[28] or that make resulting dissent likely to be epistemically detrimental.[29] If scholars can identify particular features of problematic dissent, then they could develop normative recommendations for dealing with such dissent and limiting its epistemic and social damages. We challenge this approach, however, in three main ways. We argue that reliably identifying problematic dissent, or what we will call "normatively inappropriate dissent" is an enormously difficult task, both conceptually and in practice. The first part (chapters 2–6) of the book assesses various criteria that one could use to identify normatively inappropriate dissent (NID) and concludes that they are unsuccessful. We further argue that, even if we could do so, the strategies that one could employ to address NID are not only unlikely to solve or reduce many of the problems of concern but could actually exacerbate some of them, as well as create new ones. For these reasons, we go on to argue that the focus on dissent as such might not be the best strategy when trying to deal with the problematic consequences of this dissent. In fact, focusing on problematic dissent might have the exact opposite of the desired effect, since it might distract from other epistemic and social conditions that render laypersons more susceptible to dissenting views and thereby increase the potential of NID to have detrimental effects.

In the second part (chapters 7–11) of the book we follow what, we believe, constitutes a more fruitful approach and focus on some of those epistemic and social conditions that actually contribute to making NID more damaging. We call attention to the ways in which epistemic trust is crucial to interactions between science and society, and we show how various scientific institutions and practices fail to facilitate and sustain warranted public trust. We contend that when trust is damaged, NID is more likely to have seriously adverse epistemic and social effects. We also propose several strategies that focus on facilitating trust rather than on addressing dissent as such. In our view, preventing or limiting the negative epistemic and policy consequences that NID can have requires both tackling institutional and social factors that undermine well-grounded trust and promoting strategies that contribute to ensuring the trustworthiness of scientists.

Moreover, we question an implicit assumption often present in discussions regarding problematic dissent: that public resistance to particular policies or recommendations results mainly from confusion about or ignorance of scientific claims for which there is a consensus. We show that some of the problems attributed to NID are the result of disagreements about values, not about facts.[30] Insofar as this is the case, moving policy debates forward in fruitful ways requires engaging in discussions about the values at stake, rather than the truth of particular scientific claims.

Our project thus involves both a critical and a positive approach. Our critical approach challenges a common way to frame problems with NID. Our positive proposal calls for two different strategies: promoting institutions and practices that make the scientific community more trustworthy; and recognizing the limits of science when it comes to policymaking.

Laying the Groundwork

In a book concerned with dissent about scientific claims, it is important to be clear about what to count as such. In this book, we use "dissent" in a descriptive rather than a normative way. We take dissent to be a type of disagreement, and disagreements can be reasonable or unreasonable, well grounded or not. When we refer to dissent, then, we do not intend to imply that the disagreement it expresses is necessarily legitimate. We use "scientific dissent" and similar terminology to mean *dissent about scientific claims*. That is, we are not concerned here with dissent that targets primarily political claims. The adjective *scientific*, therefore, is not meant to imply that the dissent in question is scientifically adequate dissent—simply that it constitutes a disagreement about scientific claims. Scientific dissent can thus cover a wide range of practices.[31] Various activities, such as criticizing and challenging, can count as dissent. Our use of "scientific dissent" in this work is narrower, in part following those who are concerned with cases of dissent that manufacture doubt or create controversy about consensus views.[32] We take scientific dissent to be not just any criticism about science but specifically the act or practice of challenging *a widely held scientific position*. This can include objecting to particular hypotheses, theories, or methodologies. Furthermore, scientists and their research are often the source of dissent, but in what follows we do not assume that they alone can produce dissenting views. Other relevant parties can also challenge widely held scientific conclusions, though as we will see, identifying these parties is part of the debate over what constitutes appropriate and inappropriate scientific dissent, as expertise may be relevant. Because we take dissent to be a practice and not just a position, we also include in our understanding of dissent the acts of promoting and disseminating dissenting research.

Our focus is not, however, on scientific dissent in general—the challenging of consensus positions—but on normatively inappropriate dissent. As we discuss in chapter 2, scientific dissent has various epistemic benefits that are essential to the advancement of science. Such epistemic benefits can obtain even when the dissent in question turns to be erroneous. What distinguishes NID from dissent that is merely mistaken is that it fails to yield any of the epistemic benefits that make even false dissent valuable. Hence, we take NID to be not simply dissent that advances incorrect scientific claims but, rather, *dissent that fails to promote or that hinders scientific progress*. This failure is problematic not only for obvious epistemic reasons but also for social ones, as NID can lead people to disregard well-grounded policies or oppose needed ones.

Our choice of the term "normatively inappropriate dissent" is not without importance. The cases of dissent discussed in the introduction and other similar ones have been variously referred to as "manufactured doubt,"[33] "epistemically detrimental dissent,"[34] dissent that fails to be "normatively appropriate,"[35] or dissent that artificially "creates scientific controversies" where there are none.[36] Although terms such as "manufactured doubt" are now part and parcel of both academic and popular discussions regarding what many, including us, take to be problematic instances of scientific dissent, we find the terminology limiting. To refer to cases of problematic dissent as "manufactured" or "artificially created controversies" implies that the dissent in question has been deliberately produced with negative intentions. Nefarious motives may—and indeed we believe they do—characterize problematic dissent in many cases, and we discuss this criterion in chapter 3. Nonetheless, dissent can have disastrous epistemic and social consequences even when it is created without conscious malicious intentions. Similarly, the term "epistemically detrimental" focuses attention only on one potential adverse consequence of problematic dissent—that is, it can slow or impede scientific progress. Although this might be a particularly important effect of problematic dissent, we prefer not to limit our attention to epistemic consequences alone. After all, some of the most debated concerns about problematic scientific dissent have centered on adverse social consequences such as the stalling of public policies—for example, regulations of CO_2 emissions that are arguably necessary to protect human interests and environmental well-being, or behaviors among members of the public, such as failing to vaccinate children. We believe that referring to problematic dissent as "normatively inappropriate" can capture dissent that has a variety of adverse consequences, whether or not it was created in bad faith.

Successful criteria for NID should therefore be able to reliably identify cases of scientific dissent that fail to advance or hinder knowledge production while excluding cases of dissent that are epistemically valuable (even if mistaken).[37] However, coming up with criteria for NID presents a challenge. Using the consequences of certain types of scientific dissent as an indication that it is normatively inappropriate will not do, because classifying as negative some effects of dissent assumes that the dissent in question is indeed normatively inappropriate. After all, any instance of scientific dissent, no matter how legitimate, could create confusion, promote or deter particular actions, and call into question the implementation of certain policies. In general, however, we would be hard-pressed to see those as *adverse* consequences if the dissent were legitimate. Challenging an existing scientific consensus is an inherent part of the usual process of producing scientific knowledge. Hence, confusion among the public and policymakers about the degree of existing consensus or the quality of such consensus is an adverse impact only if there is actually no real, relevant dissent in the scientific community and if the process by which consensus has been achieved is in fact reliable. Similarly, concerns about false beliefs are appropriate only if such beliefs are indeed false—that is, if they reject the existence of a consensus where there is one or its reliability where it occurs. The same follows for peoples' behaviors and support for public policies. That parents, for instance, refuse to vaccinate their children because of fears that doing so might cause autism in their children is a negative consequence of dissenting views if the safety of the vaccines in this respect is, in fact, well established. Similarly, stalling policies on mitigating the effects of global warming constitutes an adverse impact of dissent if there are no good reasons to question the scientific consensus on climate change. Thus, a successful account of NID would need to identify criteria that do not rely solely on the presumed negative consequences of the dissent in question.

Examining cases that many consider paradigmatic instances of such dissent to uncover what they have in common is similarly problematic. Whether dissent involved in denials of climate change, vaccination safety, or the relationships between HIV and AIDS are in fact cases of NID is contested.[38] Witness, for example, the debate over genetically modified products (GMPs). Some have taken dissent regarding their safety as a paradigmatic instance of NID.[39] However, although many scientists believe that the safety of GMPs is well grounded,[40] many others contend that methodological concerns such as the duration of many studies, or the limited endpoints utilized, legitimately call into question the strength of the evidence.[41] Insofar as a consensus about the safety of GMPs exists, then critics will very much disagree that their dissent is normatively inappropriate. Of course, there are also disagreements about whether a reliable consensus actually exists.[42] And if no consensus exists, then it is not clear that challenges to the safety of GMPs would count as dissent at all—as opposed to criticism or disagreements—much less as NID. Moreover, at most this strategy would allow one to *explain* why a particular instance of dissent is normatively inappropriate. However, ideally the objective would be to identify NID before it inflicts any damage, or at least when we can still minimize its negative effects. Thus we need predictive criteria and not just explanatory ones.

One possible way to deal with these challenges is to use cases of dissent that appear intuitively to constitute NID not simply as a way to uncover identifying features but, rather, as a step that would allow us to reflect on whether some of their features are inconsistent with the goals that dissent has in knowledge production. If such is the case, then this would constitute a good reason to question the value of the dissent in question. In chapter 2, then, we tackle this task and discuss the ways in which scientific dissent is epistemically and socially valuable.

Finally, while we criticize several different criteria for NID, we do not deny that NID exists or that it can constitute a problem. Neither do we exclude the possibility that some instances of dissent can correctly be identified as normatively inappropriate—that is, as dissent that fails to contribute to or that hinders knowledge production. We contend, however, that successful criteria for NID must provide predictive criteria that can *reliably* identify NID—that is, it must be able to successfully identify NID as such when the dissent in question is in fact normatively inappropriate and be able to exclude scientific dissent that is actually legitimate when such is the case.

One might be concerned that this is setting the bar rather high. If a criterion or criteria for NID could provide some reasonable guide to cases of dissent about which we ought to at least be suspicious, why should we demand more? There are, after all, many concepts that are useful even if they are not understood as corresponding to a set of necessary and sufficient conditions. For example, one might be unable to provide necessary and sufficient conditions for liberalism, but that hardly prevents us from using the term in meaningful ways. It is true that sometimes one might be incorrect in describing a certain policy as liberal when it is not or one might be wrong in declaring that some policy is not liberal when, in fact, it is. Nonetheless, people can usually correctly identify liberal policies. Why should we suppose, then, that a successful account of NID must reliably distinguish cases of NID from cases of appropriate dissent in every case?

We take it that the degree of "reliability" to which it is appropriate to hold criteria depends in part on both the aims of the criteria and the consequences of error. As we mentioned, a primary goal of identifying NID is that doing so can allow us to put strategies in place to prevent or address the adverse consequences NID can have. If we seek to identify NID in order to simply educate laypeople about the problematic nature of some dissent so that they are more attentive and less likely to be persuaded by it, then the consequences of characterizing some dissent as normatively inappropriate when it is not would not be very serious. Rough guidelines for NID might occasionally fail to capture cases of problematic dissent or accidentally pick out instances of valuable dissent, but the result would be more attentive laypersons, which would be a good thing. Toward this aim, then, criteria for NID need not be highly accurate or reliable. However, in this case, it is also not clear that criteria for NID would be very useful, since dissent against a scientific consensus should be rigorously scrutinized in any case (regardless of whether it is normatively inappropriate or not). If, however, the goals of identifying NID is to allow us to limit, discourage, suppress, or ignore certain kinds of dissenting views, then—as we will show in chapter 2—the epistemic and social consequences of mistakenly identifying some dissent as NID when it is not can be very serious. If, for example, reviewers were considering not funding research that inappropriately challenged a consensus view, or prohibiting the publication of dissenting views deemed NID, they would want to be reasonably sure that such research was, in fact, of little epistemic or social value. Otherwise, the reviewers risk suppressing dissent that could have significant benefits. In such case, it is appropriate to ask for more reliable and precise identifying criteria.

Some further clarifications regarding terminology are also in order. Throughout the book we use "scientists" and "scientific community" interchangeably, and unless otherwise noted, we do not make distinctions between different scientific communities. This is certainly a simplification, as differences do indeed exist among scientific disciplines with respect to practices, methodologies, organization, and so on. Nonetheless, we do not believe that such differences are relevant to the arguments we make here. Similarly, in much of our discussion we use "science" to refer both to the institution and the knowledge produced. However, where relevant to the arguments we present, we aim to make careful use of this distinction.

We also talk about the "public" or "laypeople," but we are quite aware that the public is not a unified entity and that laypeople do not have equal knowledge, values, interests, or backgrounds. We use the term "public" to refer to all "publics" or layperson stakeholders who might be affected by the production of knowledge, but we do so in ways that do not make the assumption that this is a monolithic group. Where necessary, we make such distinctions. Indeed, we believe that differences between stakeholders are of relevance when trying to implement some of the solutions needed to minimize the damages NID can produce.

Overview of the Book

Trying to prevent or limit the damages that can result from problematic dissent involving policy-related science might initially appear a relatively easy task. Once the dissent in question has been identified, scholars and others can attempt to discredit dissenters by showing, for instance, their financial ties to industry or political think tanks so as to call into question the quality of the dissent.[43] In an effort to correct false beliefs or reduce them, scholars could also choose to emphasize the existence of scientific consensus[44] and to limit public discussion regarding legitimate disagreements within the scientific community.[45] Scientists could also limit the influence of NID and thus reduce its presumed negative epistemic and social consequences by suppressing problematic dissenting views through the review process.[46]

But as we will see in chapter 2, scientific dissent plays essential roles in the production of scientific knowledge. First, it furthers scientific progress.[47] Dissent can contribute to limiting the influence of problematic biases of individual researchers; ensure that a full range of research projects, hypotheses, models, and explanations receive adequate attention; and provide alternative ways of conceiving phenomena. Second, even if dissenting claims turn ultimately to be incorrect, dissent may lead to new evidence in support of the consensus view, strengthen justification for the consensus, and thus increase confidence in its correctness.[48] Third, the existence of dissent can promote appropriate public trust in science. It can give the public confidence that the scientific community is open to challenges and that it duly scrutinizes consensus views.[49]

Because of the critical value that dissent has in science, any strategy intended to target dissent necessitates a reliable means to identify problematic dissent—that is, the ability to distinguish NID from legitimate and valuable instances of dissent, whether or not such legitimate instances turn out to be scientifically correct. Without the ability to reliably identify NID, an emphasis on the consensus, the silencing or restriction of dissenting views, or even attempts to simply limit opportunities for dissent can constrain valid research agendas, the pursuit of alternative areas of knowledge, and the consideration of important objections if the dissent targeted is actually epistemically valuable.[50] To the extent that such strategies might, even if inadvertently, lead to excluding the legitimate voices of those who are socially marginal, they can also contribute to social injustice.[51] Of course, insofar as education of the public and policymakers is the primary strategy used to address the adverse consequences of NID, strict identifying criteria might well be unnecessary. Education about a particular instance of scientific dissent is unlikely to do harm even if the dissent in question is normatively appropriate. Unfortunately, education is also unlikely to be very successful in addressing the challenges that NID can create.

Similarly, as mentioned earlier, the evaluation of the consequences of certain types of dissent as negative rests on the assumption that the dissent in question is indeed problematic. That people reject certain policies grounded on a presumed consensus view or that they ignore experts' advice regarding a consensus is a problem if the consensus in question is, in fact, reliable. Thus, determining whether particular consequences that stem from the existence of dissent are indeed negative, so that they can be tackled, also requires that we are able to reliably identify NID.

Chapters 3 through 5 critically evaluate specific criteria that could be used for identifying NID. We are not claiming that all those discussing or using such criteria explicitly talk about criteria for appropriate or inappropriate dissent. Nor are we arguing that the goal of the authors we discuss is to find criteria to reliably identify NID—indeed, many of them have not such goal at all.[52] Nonetheless, when discussing cases of what they argue constitute problematic dissent, various authors have pointed to particular features of the dissent they believe are indicative that the dissent in question is indeed problematic. Our goal in these chapters is to assess whether one or more of these features might be used to reliably identify NID. Some have pointed out that many instances of problematic dissent involve inappropriate intentions.[53] Could attending to the intentions of dissenters be useful in reliably identifying NID? Intuitively, dissenting research that aims to confuse the public, promote false beliefs, or manipulate policy decisions—that is, dissent that is produced in bad faith—seems unlikely to provide the kinds of epistemic benefits that make dissent valuable. In chapter 3, we assess this criterion. We explore various ways to explain why bad faith motives could indeed result in dissent that fails to promote or that in fact impedes scientific progress, and we assess their plausibility. We consider the possibility that the problem with these bad faith motives is that they insert nonepistemic values in the conduct and appraisal of scientific research and reject this option. We also contemplate whether what constitutes the problem is that bad-faith motives involve values antithetical to the epistemic aims of science. We conclude that, in spite of the intuitive appeal of attending to motivations, they cannot serve as a criterion to reliably identify NID.

Focusing on rules of engagement for fruitful discussions about competing scientific views provides another strategy for trying to reliably identify NID. Chapter 4 analyses this approach. It discusses some of the rules for effective criticism dominant in the philosophy of science literature: shared standards, uptake, and expertise. In order to advance scientific debates, some shared standards for what constitutes good evidence must exist.[54] Similarly, dissenters must engage in uptake—that is, they must take criticisms of their own views seriously and respond to these challenges by revising their arguments or explaining why the criticism is unsound or misguided.[55] Likewise, actors must have appropriate expertise in the scientific area under dispute in order to be able to propose and evaluate dissenting claims.[56] Although it would be problematic to dismiss dissent on the basis of some epistemically irrelevant feature, such as the gender, race, or political affiliation of the dissenter,[57] being an "eligible player" in the scientific-inquiry game requires that one meets certain qualifications relevant to the area of science under investigation. In chapter 4, we argue that, although these criteria appear eminently reasonable as requirements of transformative criticisms, what they actually involve is not straightforward. Indeed, we show that some of the interpretations of these criteria are likely to identify as inappropriate dissent that is actually epistemically valuable, while other interpretations of these criteria would fail to pinpoint the very cases of dissent that some consider paradigm cases of NID, such as dissent with respect to evolutionary theory, climate change, and GMO safety.

Others have called attention to the role of inductive-risk judgments in the context of problematic dissent.[58] Dissent that calls for rejecting certain consensus views related to public policy can be risky. When consensus views are mistakenly rejected, it can have serious consequences for public health and well-being. These risks may not be worth taking when the risks disproportionately fall on the public, or when the dissent in question fails to conform to widely shared standards of good science. Chapter 5 focuses on this inductive-risk account. After showing its strengths, we critically evaluate it and argue that this account also fails to provide us with a criterion to reliably identify NID. In part this is because of the difficulties presented by the criterion of shared standards in science. Moreover, because of the ambiguities present in judgments about inductive risks, the inductive-risk account also turns out to present serious problems in practice.

These chapters together provide one of the pillars of our challenge to the question of dissent: reliably identifying NID is difficult, and although some scholars have defended potential criteria implicitly or explicitly, none has been successful. Chapter 6 introduces the second pillar: even if such reliable identification were possible, it would be unlikely to help us in addressing many of the epistemic and social adverse impacts that can result from NID. Granting, for the sake of argument, that it is possible to reliably track NID, we study how this could be used to prevent or reduce the damage that such dissent can produce. For instance, once identified as NID, one could prohibit the dissent in question, target it for special scrutiny, discredit dissenters, or determine when to emphasize a consensus. Scholars could also use the ability to reliably identify NID as a way to place limits on scientists' epistemic obligations and as a source of information to guide public beliefs. We show that, although some of these strategies might have some utility in addressing some of the problems that NID can create, others are unhelpful in limiting its impacts and may even exacerbate them or generate other equally serious problems. Moreover, to ensure that only NID is classified as such, the identifying criteria would have to be correctly applied.[59] The effectiveness of the criteria would therefore also depend on the extent to which laypersons could use them in practice. This would obviously depend on how much expertise one requires to correctly employ the relevant criteria. If people must have a high level of expertise, the NID-detecting criteria would be of limited use in helping address the adverse epistemic and social consequences such dissent can create, since much of the damage is the consequence of policy choices made by laypersons.

If finding criteria to reliably identify NID is unlikely to be effective or useful, then what should we do about this dissent and its adverse consequences? Answering this question is the task of chapters 7 through 10. Here, we propose to move the debate away from dissent itself. After all, a focus on dissent is unlikely to prove particularly fruitful. Instead, we attend to factors that arguably contribute to making NID particularly damaging. We thus argue in chapter 7 that a context where the trustworthiness of scientists is called into question, and where there is an excessive reliance on scientific information when it comes to assessing policy decisions, allows NID to take hold and erroneously affect people's beliefs and ultimately their actions.

When the public fails to believe that scientific communities deserve their trust, dissenting views that contest the reliability and soundness of scientific testimony will find fertile soil. Chapters 8 and 9 focus on some of the social and institutional factors we believe cast doubt on the trustworthiness of scientists and ultimately undermine warranted trust in scientific communities and their testimony.

Chapter 8 attends to some social and institutional factors related to the practice of science that we argue contribute to the erosion of trust in scientists. Specifically, we consider the role that the increasing commercialization of science and concerns about scientific misconduct have in calling into question the trustworthiness of scientists. In a context where economic incentives and profits play a substantial role in determining the direction of research agendas, the common good of all will take a back seat. The increased commercialization of science thus undercuts one of its primary goals—that of contributing to the common good. Inadequate attention to the public good, then, contributes to undermining trust in the scientific community. Moreover, the commercialization of science results in financial conflicts of interest for both scientists and research institutions, and such conflicts threaten the integrity of research. Conflicts of interest, although not necessarily resulting in biased science, certainly create a fertile ground for it. Given that in general it is difficult for laypersons to assess whether or not conflicts have adversely affected the justification for particular scientific claims, rampant conflicts of interest cast doubts on the trustworthiness of scientists. We likewise consider the phenomenon of scientific misconduct. We show that scientific misconduct, particularly when understood as including questionable research practices, is far from uncommon, and that policies to prevent, detect, investigate, and sanction such misconduct are either nonexistent or inadequate, meaning that the current research system is probably contributing to the presence of cases of misconduct. Unsurprisingly, we claim that this phenomenon also undermines trust in scientific communities and their claims. We consider some strategies for changing these trust-undermining practices so as to better facilitate and maintain rational trust in scientific communities.

Chapter 9 deals with another factor that plays a role in eroding trust in science: concerns about the negative influence of nonepistemic values in science. The use of social, political, or personal values and interest can threaten the epistemic integrity of the research. If the public believes that scientists have a political agenda that biases their research, this can call into question the trustworthiness of scientists in a particular research area and thus can undermine public trust in science. To address this source of public mistrust, it might be tempting to encourage scientists to adhere more strictly to a value-free ideal of science so as to ensure that nonepistemic values do not influence scientific reasoning. This, however would be a mistake. Notwithstanding numerous examples from the history of science where the use of nonepistemic values have led to bias, such values also play necessary and important roles in scientific reasoning. Facilitating warranted trust, we argue, requires that we develop and implement strategies likely to reduce the illegitimate use of nonepistemic values in science. Institutional practices, such as promoting public avenues for criticism or promoting diversity among scientists, increase the opportunities for identifying and critically evaluating values, as well as for correcting the biasing influences such values may have. Such practices are likely to be more successful both in ensuring the epistemic integrity of the research and in creating and sustaining justified public trust in scientific communities. These sorts of mechanisms, however, fail to address reasonable concerns the public may have about the priority of certain values over others. Such concerns can also undermine justified trust in scientific communities. Even if nonepistemic values do not undermine and can actually improve the epistemic soundness of research, leaving these decisions in the hands of scientists alone is a different matter altogether. Scientists have no expertise in this area and their value choices are unlikely to be representative of the diverse interests that exist among stakeholders. Thus, to create and sustain warranted public trust, the scientific community needs to implement strategies that promote transparency of values and are inclusive and representative of the diversity of stakeholders' interests.

If we are correct in our analysis, then encouraging institutional and social practices that make the scientific community more trustworthy is likely to be a more effective strategy to address the problems that NID can create. This will require significant empirical research aimed at identifying the kinds of mechanisms that can be successful in doing so. Of course, even if such mechanisms can be identified, we recognize that implementing them will not be an easy task. But, as the continuing debate on global warming shows, attempts to target some dissenting views so as to avoid their negative social and epistemic consequences is not an easy task, either. Moreover, promoting institutional changes that encourage warranted trust in the scientific community is likely to have additional benefits that have nothing to do with correcting or limiting the problems that NID can produce.

In chapter 10, we change gears and propose a second recommendation to deal with the negative adverse effects that NID can have: we must recognize the limits of scientific evidence when it comes to public policymaking. All the prior chapters accept an assumption that underpins much discussion regarding problematic dissent: that resistance to particular policies and actions stems in great part from the presence of such dissent. The contention is that failures to, for example, support emissions regulations, vaccinate children, or embrace widespread use of GMOs are to a large extent the result of dissenting views that end up confusing the public and policymakers about the state of the science. Chapter 10 interrogates that assumption. We argue that resistance to science-based policy recommendations can arise from disagreements about the values that underlie certain policy choices, rather than simply from confusion about or ignorance of the state of the science. In this chapter, we call attention to an important aspect of policy-relevant science that has been obscured in discussions over NID: the role of values in policy decisions. We identify different ways in which value disagreements can reasonably lead people to conflicting views about whether to reject or support particular policies or actions. Specifically, we argue that failure to accept certain policies or actions may be the result of disagreements about what has value, how to interpret particular values that are shared, how to weigh competing goods when they conflict, or how best to promote particular values or policy goals. Insofar as resistance to certain policies is the result of disagreements about values rather than simply the result of disagreements about the scientific evidence, understanding the ways in which differences in values can influence policy debates is crucial to being able to advance such debates in fruitful ways. As important as science is for sound public policy, at least in the context of some public policies, a focus on the science can take us only so far. Finding ways to solve value disagreements is thus likely to help address the negative effects that NID can have.

How this Book Is Different

There are genuine concerns about the gap in various scientific areas between what scientists take to be a matter of consensus and what laypersons often believe regarding such consensus. Many have argued that this gap results in significant part from the presence of problematic dissent, or what we here call NID. In this book, we challenge this approach to the problem. Without denying that some dissent can fail to advance or even hinder knowledge production, and that its existence can have adverse consequences such as public confusion, stalled policies, wasted resources, and intimidation of scientists, we believe that the central problem has been misdiagnosed.

This book aims to consider four central questions: Is NID a primary culprit of these various problems? Insofar as NID *is* a problem, is there a way to *reliably* identify it? If we were able to reliably identify such NID, what could we do about it? And if reliable identification is not possible, what else can be done to minimize the problems of concern?

Other authors have assumed, either implicitly or explicitly, that NID is indeed a primary culprit, and they have pointed out features that could be used to identify such dissent. Presumably, once identified, the dissent in question could be managed in some way by, for example, prohibiting it, ignoring it, or exposing it.[60] This book makes a significant contribution to the discussion by providing a systematic assessment of these varied features and by showing that they are not as straightforward or successful as those using them have assumed. Indeed, we show that, even if successful, efforts to provide criteria to reliably identify NID would fail to address the negative consequences that can result from it.

But the contribution of this work goes further. We reframe the discussion in a more fruitful way to focus not on dissent itself but, instead, on the social and institutional conditions in which NID arises—conditions that allow such dissent to have adverse effects. In particular, NID is more likely to be damaging when warranted public trust in scientists falters and when we fail to recognize the limits of scientific evidence when making policy decisions. This reframing sheds light on various features of scientific practices that cast doubt on scientists' trustworthiness and contribute to undermining warranted trust. Similarly, it also calls attention to the relevance of inescapable value judgments that play a role in policy decisions where scientific knowledge is also important. This reframing is particularly important because much of the discussion on NID has implicitly or explicitly assumed that people's rejection of the reliability or existence of a genuine scientific consensus regarding, for instance, climate change or vaccine safety stems from the influence of such dissent. Under this assumption, promoting strategies to better communicate to the public and policymakers that a consensus exists seem appropriate. And this is, in fact, what many of those engaging in this debate have done.[61] But if a significant reason for why NID is so effective is that laypersons question the trustworthiness of scientists, or that they have different value assumptions, then the success of this strategy will be limited. Insisting on the existence of a consensus will have little impact on those who doubt the very foundations of that consensus or who disagree with them because they hold conflicting values. Furthermore, insofar as these strategies obscure what are at least important contributory factors, they prevent us from identifying more successful alternatives.

This book advances discussions within the philosophy of science by synthesizing and drawing connections from several distinct bodies of literature—specifically, literature on consensus and dissent, the roles of values in knowledge production, the relevance of values for science-based public policy, and epistemic trust. We aim to show that attempts to understand and address concerns related to dissent and public policy can be greatly informed by elucidating the roles—both negative and positive—that contextual values play in science and in policymaking. Other discussions of problematic dissent have been particularly concerned with the negative role that values can play, leading to conflicts of interest and bias,[62] or do not attend to the role of such values altogether.[63] At the same time, many of those writing on values in science have not been attentive enough to the ways in which the presence of contextual values may provide fertile ground for problematic dissent that can lead to negative consequences.[64] We thus advance current debates by underscoring the limits of scientific knowledge when trying to solve controversial policy decisions. Likewise, questions about trust, both moral and epistemic, are a relatively new concern in philosophy of science.[65] Though a concern for social scientists, few philosophers of science have explored the role of trust and trustworthiness in relation to current debates regarding climate change, vaccine denials, or criticism of GMO.[66] We advance these discussions by calling attention to the role that trust—or lack thereof—plays in current disputes over scientific dissent, as well as to its importance when trying to address the problems that NID can produce.

This book also differs from the usual way in which many philosophers of science approach scientific practices. Unlike some accounts,[67] we do not aim to offer an idealized model of scientific practices.[68] Our analysis is tied to the context in which science is currently practiced: one that is largely capitalist and increasingly driven by commercial interests, where participants do not have equal access to education or expertise, nor equal power to make their voices heard. Insofar as we are concerned with diagnosing very real problems related to dissent and the negative consequences that impact society, we believe that an analysis both of the problem and of the solutions must be attentive to these conditions that are so difficult and slow to change.

The scope of our project is also limited in important ways. We do not attempt to provide an exhaustive analysis of all the factors that contribute to individuals' resistance to particular beliefs or policies. Social scientists are doing important work on these issues, including the ways in which such resistance may be caused by cognitive biases, political orientation, social identity, or moral worldview.[69] While we aim to be attentive to this empirical evidence in our arguments, it is beyond the scope of our project to provide a full analysis of belief formation or policy preferences. Rather, our goal is to evaluate some contextual aspects that arguably make some dissent particularly problematic.

Similarly, we do not aim to develop or defend a particular account of trust in general or epistemic trust in particular. Instead, we hope to show how questions about trust, moral and epistemic, bear on concerns that have arisen in relation to people's perceptions of scientific dissent and consensus. Nor do we provide an exhaustive examination of all the variables that influence public trust in scientific communities or all the factors that impact scientists' trustworthiness and people's assessment of it. However, we believe that the factors we address are particularly relevant to understanding concerns that dissenting views in various policy-relevant scientific issues give rise to, as well as proposing effective responses to them. Similarly, although we consider and evaluate several strategies that could be effective in facilitating and sustaining warranted public trust in science, we do not claim to offer an exhaustive list of potential solutions. Our aim is to show that reframing debates about dissent can reveal several important future areas of inquiry that will be important to enabling and nourishing justified public trust in scientific communities and to moving policy debates forward.

We do not attempt to offer an exhaustive account of the many ways in which contextual values may play a role in policy decisions or how such values inform people's beliefs. Nor do we claim that the options we discuss to attempt to solve value disagreements in science-based policy decisions are the best or only ones. Our goal is simply to call attention to the limits of scientific knowledge when dealing with complex and controversial policy issues and to encourage scholarly work that can provide strategies to address the effect that value disagreements have on policy decisions.

Finally, it is worth emphasizing that although our arguments ultimately show the difficulties of identifying NID, this should not be understood as a defense of many of the instances of problematic dissent we discuss here. We do believe that it is reasonable to think of at least some of the dissenting views we present as failing to provide any epistemic benefit. But this recognition alone is unlikely to help us either prevent or solve the epistemic and social problems NID can create. Thus, if our goal is to do something about dissent that is identified as normatively inappropriate, we need some reliable criteria. Our concern, however, is that providing such criteria is difficult and risky, as well as unlikely to be effective in helping us address the damage such dissent can inflict.

Finding criteria to reliably identify NID is challenging precisely because dissent plays a crucial role in the production of scientific knowledge. Indeed, the valuable role that dissent plays means that scientific communities have an obligation not simply to tolerate but also to actively seek out and engage with dissenting views. Any attempt to tackle NID must recognize and safeguard the epistemic and social value of dissent.

2

The Important Roles of Dissent

For much of the twentieth century a consensus existed within the medical community that the cause of stomach ulcers was an excess of stomach acid and dietary factors. Barry Marshall, an internist, and Robin Warren, a pathologist, first reported "unidentified curved bacilli" on gastric epithelium in active chronic gastritis in two letters published in the *Lancet* journal in 1983,[1] and later postulated that the bacteria could possibly play a role in peptic ulcer disease.[2] They argued that the bacteria could not be seen with the usual staining methods employed at the time, which explained why the bacteria had mostly been overlooked. The possibility that bacteria could be a cause of stomach ulcers was met with significant skepticism, in part because it was inconsistent with the prevailing view that bacteria could not survive in the highly acidic environment of the stomach.[3] A decade later, the National Institutes of Health Consensus Development Conference concluded that there was a strong association between *Helicobacter pylori* and ulcer disease, and recommended that ulcer patients with *H. pylori* infection be treated with antibiotics.[4] Medical practice was slow to follow recommendations. Studies showed that by the year 2000 the overwhelming majority of ulcer patients were still treated primarily with anti-secretory medications, with only 17% of them receiving antibiotic therapy.[5] The dissenting research by Marshall and Warren, however, eventually prevailed, improving the treatment for peptic ulcers and earning the Nobel Prize in Physiology in 2005.

There is widespread agreement that dissent plays a crucial role in the generation of scientific knowledge.[6] Even those who believe that consensus is the ultimate goal of scientific inquiry take dissent to be important to achieving or maintaining the integrity of that consensus.[7] As we will see, dealing with normatively inappropriate dissent (NID) is challenging precisely because dissent is critically valuable in the production of knowledge. The purpose of this chapter is to provide an overview of the ways in which dissent from a scientific consensus is epistemically valuable.

Dissent furthers scientific progress. It can do so in several interrelated ways, including correcting false empirical assumptions, providing alternative ways of conceiving phenomena, and challenging value judgments. Dissent can also strengthen the justification for consensus views and thus increase confidence in their correctness. Consensus views are more likely to be reliable when they survive critical scrutiny than if they go unchallenged. Additionally, the existence of dissent, even when it is incorrect, can foster warranted public trust in science. That is, it can assure the public that scientific inquiry is an open and critical process whereby scientists consider evidence and challenges carefully. In the last section of this chapter, we argue that because dissent yields these benefits, it imposes epistemic obligations on scientific communities. We discuss what such obligations are and show that fulfilling them can also result in important benefits.

Understanding why dissent is valuable is relevant, for several reasons. It helps us to identify the aspects of dissent that need protection in order to foster scientific progress. In addition, attending to what is valuable about dissent is necessary for distinguishing between normatively appropriate and inappropriate dissent. If some dissent has features that are incompatible with the epistemic roles that dissent plays—that is, if there is dissent that either fails to contribute to knowledge production or actively hinders it—then we have good reasons to characterize such dissent as normatively inappropriate. Insofar as this is the case, then, we could have a justification to impose limits on such dissent without fearing negative impacts on the production of scientific knowledge.

The Value of Scientific Dissent

In *On Liberty*, John Stuart Mill famously defended the importance of dissent and its expression in the political sphere.[8] He argued that dissent and its dissemination are essential not only to promoting truth but also to ensuring that the truths held are done so for the right reasons. In his words:

> [T]he peculiar evil of silencing the expression of an opinion is that it is robbing the human race, posterity as well as the existing generation—those who dissent from the opinion, still more than those who hold it. If the opinion is right, they are deprived of the opportunity of exchanging error for truth; if wrong, they lose, what is almost as great a benefit, the clearer perception and livelier impression of truth produced by its collision with error.[9]

Dissenting views, however implausible they might first seem, may in fact by true. Even when they are not true, they provide us with increased understanding of why we hold the views we do.

Dissent in the scientific sphere is valuable for similar reasons. Clearly, insofar as some accepted scientific claims are false and dissenting views true, censoring or actively ignoring dissenting views will prevent the emergence of truth. Scientists, like all epistemic agents, are fallible, after all. A cursory review of any history of science book will show multiple instances in which a view, or some aspect of it, that enjoyed widespread consensus within a scientific community later turned out to be false. The geocentric view of the universe, the phlogiston theory, the fixity of species, the immobility of the continents, phrenology, and the existence of ether all were ultimately abandoned because of persistent dissent from the prevailing consensus. Science often gets it wrong, even when there appears to be overwhelming evidence and widespread agreement. Indeed, that science "gets it wrong" is arguably intrinsic to the nature of scientific progress. The common phenomenon of underdetermination of theory by evidence and human fallibility are sources of error. The fact that scientific consensus has often been mistaken in the past gives us good reason to think it is likely to be mistaken again in the future, and thus good reason to value dissenting views.

Scientific reasoning has the potential to go wrong in many ways. Scientific hypothesis or theories are not tested in isolation; they require a host of background assumptions in order to make empirical predictions, employ certain conceptual frameworks, justify particular methodologies, classify data, and make evidential inferences about the data in question.[10] These background assumptions or auxiliary hypotheses are often implicit and not uncommonly adopted by scientists unconsciously. But, because they are frequently unconsciously assumed, even thorough and well-trained scientists sometimes rely on background assumptions that are actually false. Obviously, it is generally easy to see mistaken assumptions in hindsight, but such assumptions can be difficult to identify at the time, particularly if they are widely shared or supported by the available evidence. Similarly, inappropriate methodologies, faulty evidence gathering, and incorrect interpretation of data can result in error.

Dissent can thus contribute to promoting scientific truths and scientific progress in various interrelated ways. First, it can challenge incorrect empirical assumptions.[11] The case of Marshall and Warren's work on H. plylori discussed at the beginning of this chapter exemplifies this benefit of dissent. At the time, their claims seemed extraordinary—and extraordinarily wrong—both because no such bacteria had been observed previously and because the scientific community accepted the belief that bacteria could not survive in highly acidic environments. Marshall and Warren's dissenting research, however, clearly played an important role in correcting those false background assumptions, advancing medical knowledge, and improving the treatment for peptic ulcers.

The work of cytogeneticist Barbara McClintock also illustrates the importance of dissent in challenging problematic empirical assumptions.[12]

Her maize breeding experiments provided the first evidence of transposable elements—also known as "jumping genes"—in the genome.[13] These transposable elements are DNA sequences that move from one location on the genome to another.[14] Her work thus showed that genomes were not stationary entities but, rather, were subject to alterations and rearrangements, and that genomic replication did not always follow a consistent pattern. She also found that, depending on where these mobile elements were inserted into a chromosome, they could reversibly alter the expression of other genes. Such suggestions directly contested the neo-Darwinian paradigm that reigned in biology during the 1950s, and that maintained that the environment could not influence the genomes other than by random mutations.[15] Yet while this paradigm was widely accepted at the time, McClintock's work revealed that it rested on several problematic assumptions and thus her research contributed to significant progress in biology.

A second way that dissent can further scientific progress is by providing new ways of conceiving of certain phenomena and with that, offering novel opportunities to tackle problems in productive ways.[16] Dissenting research is likely to result from the use of new methodologies, novel research questions, and innovative explanations, often because these reject certain assumptions of the consensus view. Dissenting views thus can make room for alternative possibilities that are obscured for those who hold the majority view.

McClintock's work with maize also provides an example of this benefit. Gregor Mendel used maize to corroborate his experiments with peas. Nonetheless, maize did not become an important model organism until the 1920s and 1930s, when a group of geneticists at Cornell University, which included McClintock,[17] began to work with it. Her use of maize as an experimental system for genetics was uncommon at the time for molecular geneticists, who usually made use of bacterial systems.[18] McClintock's research on transposable elements exploited particular aspects of maize that would have been more difficult to do with model bacterial organisms.[19] Thus, in working with an unusual model organism, she opened new possibilities for experimenting on genes.

Similarly, work in archeology regarding the evolution of human tool use provides an illustration of how dissenting views can promote alternative ways to look at evidence. Historically, many archeologists focused solely on the activities traditionally assigned to males, such as hunting.[20] As a result, scientists came to agree that human tool use emerged primarily from the activities of male hunters. This consensus was challenged in the 1970s and 1980s, when more females entered the field of archeology and began asking new questions about women's activities.[21] These new questions led female archeologists to reveal new lines of evidence: baskets and reeds used for foraging could also be thought of *as* tools, thus contesting the hypothesis that the evolution of human tools was primarily the result of men's activities.[22] Willingness to raise new questions challenging the consensus view ultimately enriched our understanding of the evolution of tool use and provided new areas of investigation within archeology.

In addition to challenging assumptions and providing alternatives, dissent can advance scientific knowledge by calling attention to and disputing value judgments at stake in scientific reasoning. There is a growing consensus among philosophers of science, science studies scholars, and even scientists that ethical and social values, or contextual values, play a role in scientific knowledge production in a variety of ways.[23] Such values may influence the framing of research problems;[24] they can affect the use of particular ontologies, models, or conceptual categories;[25] and can impact the selection of methodologies and the assessment of risks.[26] Many times, value judgments that influence these decisions are implicit and unconscious. As a result, these judgments can alternatively enhance or obscure knowledge of phenomena, depending on the values at stake and who shares them. Dissenting views that call attention to and challenge the use of certain value judgments in science can thus promote scientific progress.

For instance, value judgments are at stake in the choice of methodology that scientists use to study the toxicity of insect-resistant maize. The genetically modified (GM) plant contains an inserted gene from the bacterium *Bacillus thuringiensis* (Bt) that produces the Bt toxin. This endows the resulting GM Bt maize plants with a resistance to certain pest species. The established methodological norm in toxicity studies is to use the purified Bt protein from the bacteria rather than using purified plant protein.[27] This methodological choice is grounded on the assumption that transgene insertion and integration are sufficiently precise and well-controlled processes. It also presupposes that the creation, from such processes, of other products that might have different side effects than the ones produced by the purified Bt protein is unlikely.[28] The plausibility of these assumptions rests on value judgments about what counts as "sufficiently" precise and well controlled so as to be safe, or how *bad* the risks of different side effects might be, given the likely benefits of such crops. Questioning these value judgments has led some researchers to use purified plant protein to assess Bt toxicity, which has produced dissenting findings on the safety of GM crops.[29] That is, they are using alternative methodologies that may yield different evidence because they challenge the value judgments implicitly supporting the consensus view. At the very least, this presents a larger range of possibilities regarding the impacts of GM crops and increases our understanding of them.

Some dissent with respect to modeling climate change also has advanced scientific progress in this way. Historically, the standard practice for modeling climate impacts was to measure the aggregate effects of factors that were easily quantifiable, such as crop yield, loss of life, dollars lost in GNP, or destroyed property.[30] This methodological practice was grounded on various value judgments, such as the assumption that the distribution of impacts is not important, or that effects difficult to measure should be disregarded even though they might be relevant to some stakeholders.[31] But the use of such value judgments in the construction of climate models had important consequences. For example, many integrative assessment models measured the aggregate impact of climate change on food production, as opposed to examining impacts at a regional or local level.[32] Such models suggested that although agriculture would decrease in places where droughts or rising sea levels caused land to become unusable, this would be offset by increased food production in areas that would have milder climate conditions.[33] Dissenters attentive to social justice concerns objected that measuring the aggregative effects of climate change on food production obscured how access to food may be affected in ways that reinforce or exacerbate existing social inequalities.[34] They called attention to the fact that, while the models indicated that the world's food supply may not be drastically reduced, they failed to account for the particular effects on the Global South areas, which already experience food scarcity. For dissenters, this was particularly problematic in a context where such countries arguably have less responsibility for anthropogenic climate change. Dissenters then challenged existing modeling practices and proposed new ones that involved information both about the aggregate impacts expected from climate change and about the distribution of those impacts to ensure that costs and benefits could be distributed equitably.[35] As a result, climate science advanced so as to produce data relevant to developing just policies and adaptation priorities.

Yet, dissent is valuable in other ways besides advancing scientific progress. As Mill's quote earlier indicates, dissenting views are epistemically valuable even when they turn out to ultimately be erroneous. Even if a consensus view holds up over time under significant scrutiny, the challenge can sharpen our understanding of particular phenomena, can help generate new evidence or reasons for believing that the consensus view is correct, and can allow scientists and the public to have more confidence that the established consensus is more than mere dogma. Consider, for example, concerns about a possible link between the measles, mumps, and rubella (MMR) and thimerosal-containing vaccines such as the diphtheria, tetanus, pertussis (DPT or DT) and the development of autism spectrum disorders.[36] Challenges to the consensus about the safety of childhood vaccines have led to research aimed at directly assessing the existence of such a link.[37] Thus, there is now new and extensive evidence that supports the consensus view that childhood vaccines do not cause autism. Although the dissent was incorrect, it led to new research that has resulted in more reliable evidence.

Finally, scientific dissent, whether correct or not, can be valuable not only for epistemic reasons but also for social ones. For instance, and particularly when dealing with policy-relevant science, dissent can strengthen warranted public trust in an existent scientific consensus and thus motivate relevant public actions. When the scientific community is open and welcoming of dissenting views, it signifies that the consensus has been reached through a rigorous and fair process.[38] When a consensus goes unchallenged, the public might worry that the consensus in question is unreliable and that it is not receiving the sort of scrutiny that it should.[39] This can be epistemically problematic insofar as potentially false assumptions might go uncorrected. But it can also be socially problematic. A consensus view that goes unchallenged, even if correct, may have the unintended consequence of *decreasing* public trust in science. The existence of dissent can reassure the public that the scientific community is appropriately scrutinizing the research, that external interests are not biasing the results, and that consensus that persists in the face of dissent is ultimately well grounded. Indeed, arguably this is one reason the consensus on climate change is impressive. The central consensus view remains well supported despite many rigorous attempts to challenge it.

Taking the Value of Dissent Seriously

The valuable roles that dissent can play call for scientific communities not merely to *tolerate* or be open to dissent when it arises but also to actively *seek* and *engage* dissenting views. In particular, the crucial role of dissent in promoting scientific progress imposes epistemic obligations on scientific communities to: provide public venues for dissent; encourage and secure the participation of dissenting views originating from a diverse community of experts and stakeholders; engage in uptake of dissent; and ensure that dissenting views are not dismissed on the basis of epistemically irrelevant factors.[40]

Dissent can only yield epistemic and social benefits if there are mechanisms in place to solicit and disseminate dissenting views. Peer review is one way to fulfill the obligation to provide public venues for scrutiny of consensus views. Although far from perfect,[41] the process is useful in checking the quality of submitted work to grant-awarding agencies, journals, or conferences—that is, that the data seem right, that the conclusions are well reasoned—and in improving the quality of the research itself. Reviewers may point out prior research important to the work under consideration, point out flaws or limitations in the methodologies researchers have used, request new experiments, call attention to errors in the interpretation of the data, and—particularly important— bring to bear a different point of view on the phenomena.

Unless open to diverse points of view, the peer review process will fail to reap the benefits of dissent. Dissenting views will be unlikely to be given a fair chance if peer review draws mainly on experts within a particular field or discipline where scientists share relevant assumptions in virtue of common training and practices. Thus, in order for dissent to effectively play its valuable role in revising incorrect beliefs, scientific communities must also strive to involve diverse participants.[42] To be sure, what sort of diversity is relevant to producing dissenting views that will yield epistemic benefits depends on various aspects of the research in question and the scientific community. In some cases, the inclusion of participants with a diversity of values and interests will be beneficial in identifying and scrutinizing background assumptions.[43] Similarly, ensuring the participation of people from different social positions or those likely to have different kinds of life experiences can also be important to ensure the objectivity of science.[44] Diversity of ideas can also contribute to the kind of creativity needed to foster alternatives.[45] In some areas of research, diversity of methodology and disciplinary expertise will be of particular relevance.[46] While it is an open question what type of diversity will best contribute to producing and nurturing epistemically beneficial dissenting views, it is clear that some such diversity will be vital to do so.

The importance of dissent in promoting scientific progress also imposes obligations on members of the scientific community to listen and take the dissent in question seriously; that is, if dissent is to yield epistemic benefits, the scientific community must engage in what Helen Longino refers to as "uptake."[47] The obligation to engage in uptake does not require that proponents of the consensus accept the claims of dissenters. Rather, it entails a *prima facie* duty on the side of those who support the consensus to respond to the dissenting evidence and arguments, either by refining or revising the consensus views or assumptions or by explaining why the dissent is misguided or inadequate. Clearly, dissent that is ignored by relevant members of the scientific community will have little effect on the improvement of knowledge production. Even if dissent that turns out to be correct is eventually given the attention it deserves, failing to engage with dissenting views when they arise can slow scientific progress and delay benefits to societies. Consider the case of Ignaz Semmelweis, a nineteenth-century Hungarian obstetrician who pioneered antiseptic procedures.[48] Puerperal fever, or postpartum infection, was common in hospitals during his time and fatal in high numbers. While practicing at the Vienna General Hospital, Semmelweis noticed a high death rate among women who delivered their babies with the assistance of doctors and medical students, whereas women who were assisted by midwives had a much lower death rate. After considering several hypotheses for the discrepancy in mortality rates, he concluded that the higher death rates for the women treated by doctors and medical students were associated with the latter's role in the handling of corpses during autopsies and before attending the pregnant women. The midwives, on the other hand, did not partake in autopsies. Semmelweis associated the exposure to decomposing organic matter with an increased risk of puerperal fever, and determined that washing hands and cleaning medical instruments might reduce the mortality. He then instituted a mandatory handwashing policy in a chlorinated lime solution for medical students and physicians before and after attending patients. The mortality rate fell to below 1%, similar to the one for women attended by midwifes. In spite of the success, the medical community did not give Semmelweis's work appropriate attention, and thus it was slow to institute his suggested modification, with adverse consequences for the lives of many women.[49]

Certainly, that some dissent is ignored by relevant scientific communities need not be the result of irrationality or bias or a complete lack of uptake. As it seems to have been the case with Semmelwies's hypothesis,[50] various factors can explain why particular dissenting views were disregarded when proposed. Lack of instrumentation to test a hypothesis, reasonable commitments to particular established theories, underdeveloped methodologies, and institutional considerations—all can lead the scientists to disregard instances of dissent. Be that as it may, given the value that dissenting views can have, arguably scientific communities have obligations to take such views seriously so as to determine their importance in fair ways.

The role that dissent plays in advancing knowledge production similarly requires that scientific communities grant dissenters what Longino has called "equality of intellectual authority."[51] Scientific communities must treat relevant members as equally capable of offering persuasive and conclusive reasons, even when such reasons challenge a consensus view. Clearly, they ought not ignore dissenters simply because of irrelevant characteristics such as gender, ethnicity, or nationality, as such exclusion will impede legitimate dissenting views from consideration or delay the benefits of dissent. The reception of Vera Rubin's research illustrates the drawbacks of failing to grant relevant members of the scientific community equality of intellectual authority. Vera Rubin was one of the first cosmologists to posit the existence of dark matter to explain the rotation of galaxies,[52] and to argue that there is a universal galaxy rotation curve.[53] In her initial work on galaxy rotation, she discovered that galaxies were not, as it was thought at the time, evenly spaced throughout the universe but, rather, that they formed clusters, and that galaxies and clusters orbited around a central point.[54] When Rubin presented her initial work on galaxy rotation to the American Astronomical Society, it was dismissed, at least in part because she was a female scientist in 1951 presenting a view that dissented from the consensus view. Although her abstract was published with all the other abstracts presented at the conference, the paper itself was rejected for publication by both the *Astrophysical Journal* and the *Astronomical Journal*. The scientific community also received her Ph.D. dissertation work on galaxy cluster rotation with skepticism and the *Astrophysical Journal* also rejected it for publication.[55] It was then published in the *Proceedings of the National Academy of Sciences*.[56] In spite of its publication, Rubin's work on a universal galaxy rotation curve did not receive much attention from the astronomy community until decades later.

Social categories such as gender, ethnicity, or class are inappropriate reasons to exclude individuals who are otherwise appropriately qualified. Indeed, if particular social categories are relevant to the acquisition of knowledge, excluding people on those grounds is epistemically unwise.[57] Granted, there are significant questions about who the relevant members of a scientific community are, whether only those who are recognized experts within a field deserve to be granted equal intellectual authority, and *who* should be considered an expert. Longino has advocated for a "tempered equality of intellectual authority,"[58] where participants are granted intellectual authority when they have training, experience, or knowledge that bears on the research in question. This would certainly include trained scientists, but may in some cases include also nonscientist stakeholders. We will return to these issues in detail in chapter 4. For now, it is sufficient to acknowledge that for dissent to be effective, the scientific community must *prima facie* recognize and respect the epistemic authority of the dissenters.

Fulfilling these obligations can also have important social benefits. A scientific community that embraces and nurtures dissent is less likely to exclude the voices of those who are socially marginal and thus less likely to commit epistemic injustices.[59] These voices, which belong precisely to those who lack access to the formal resources often needed to participate in scientific debates, have been historically underrepresented in the scientific establishment. Insofar as these groups usually have different experiences, background assumptions, and interests than those of scientific communities, they are likely to hold assumptions that differ from the majority communities and thus to form dissenting views. This would obviously provide epistemic benefits. But encouraging and engaging dissenting views can bring the voices and interests of often marginalized groups to the forefront and can thus also promote social justice.

Indeed, there is growing recognition that securing the participation of and critical feedback from diverse stakeholders in policy-related science can have not only epistemic but also social benefits, precisely by giving a voice to those who are often voiceless.[60] Take, for instance, the changes in the design of drug trials that resulted from the dissenting views of gay AIDS activists during the 1980s.[61] These well-informed activists challenged the randomized clinical trials orthodoxy and questioned, for example, the exclusion of subjects with co-morbidities, prohibitions against taking concomitant medications, and requirements that subjects avoid participating in multiple trials. In challenging assumptions about the randomized clinical trial standard, gay activists forced the research establishment to design clinical trials that could serve AIDS patients, most of whom suffered from multiple health problems that needed simultaneous treatments, and gained the inclusion of a more diverse subject population in antiretroviral trials.[62] AIDS clinical trials thus increased their external validity. But activists not only were able to achieve changes in the conduct of clinical trials but also gained intellectual authority and respect from the scientific community—at a time when discrimination against homosexuals was rampant—and ultimately became members of the National Institutes of Health's (NIH) AIDS Clinical Trials Group.[63]

Moreover, as this example also illustrates, securing dissent from marginalized groups increases the chances that resulting knowledge and interventions will benefit such groups, as opposed to no one or the privileged few. Given the power of science to affect human well-being—through new medical interventions, epidemiological knowledge, understanding of social practices, and so on—limiting the exclusion of marginalized groups in science is of no small importance when trying to promote social justice. If knowledge production is likely to serve the interests of some stakeholders rather than those of others, then dissent can be valuable in promoting scientific knowledge that attends to the interest of all.[64] The studies conducted on dioxin to understand the potential toxic effects of Agent Orange—an herbicide widely used during the Vietnam War—illustrate this point. Communities in Vietnam, where heavy spraying of Agent Orange occurred, noticed high rates of a variety of birth defects and serious health problems.[65] Up through the 1990s, the scientific consensus in the United States was that TCDD, a type of dioxin used in Agent Orange, did not cause birth defects because it could not bind with or alter the structure of DNA.[66] Moreover, initial studies done in the United States produced little evidence to support a link between Agent Orange and the kinds of birth defects and health problems existing in Vietnam.[67] Nevertheless, as Nancy McHugh has shown,[68] the studies conducted in the United States involved preclinical trials with animals or randomized controlled clinical trials with U.S. male veterans who had been exposed for relatively short periods of time and through direct contact with the agent. People living in Vietnam, however, had prolonged exposure to the dioxins in the soil, ground water, and food.[69] Researchers collaborating with people residing in several communities in Vietnam uncovered that residents' practices and living conditions increased exposure to contaminated substances in ways not experienced by U.S. veterans.[70] Women, in particular, were exposed to higher risks in part because of differences in deposits of fatty tissue where dioxin is stored. Moreover, dioxin was in their breast milk.[71] Activism by members of affected communities led to new lines of research, particularly within Vietnam, that more directly attended to exposure variables and paths that people living in those conditions were likely to experience.[72] Activists' work contesting the results on dioxin effects effectively led investigators to conduct research that addressed the communities' concerns. The initial failure to engage with local communities in Vietnam resulted both in epistemically misleading studies about the toxic effects of TCDD in Vietnamese populations and in the dismissal of a legal suit against the U.S. government and chemical companies—compensation that could have helped address the health problems they continued to face.[73] The suit was dismissed because the judge found that there was insufficient evidence at the time that TCDD caused the sorts of birth defects and health problems claimed by the Vietnamese communities. Yet this lack of evidence resulted precisely from a failure to conduct research attentive to the kind of exposure experienced under the conditions in which these communities lived.[74]

Marginalized groups are also more likely to challenge oppressive assumptions and values, such as sexist and racist ones, and thus can contribute to the creation of knowledge that reduces such oppression.[75] For instance, Sarah Blaffer Hrdy has shown the ways in which female primatologists in the 1980s contested evolutionary models of sexual selection.[76] Such models portrayed females as highly selective and sexually "coy," and males as undiscriminating and sexually promiscuous.[77] Several assumptions grounded the sexual selection theories that the scientific community accepted: that male investment in offspring production was small relative to females, that males had greater variance in reproductive success than females, and that the only rationale for females to mate was fertilization.[78] Evolutionary biologists accepted these assumptions uncritically, at least in part because they were consistent with predominant gender norms and stereotypes that represented women as sexually passive, choosy, and monogamous and men as sexually aggressive and promiscuous. However, as female primatologists began entering the field, they asked about the work of natural selection on females and not just males, and they began to question the assumptions underlying sexual selection models. They also began finding evidence that claims about lower investment in offspring by males depended on controversial measures, that different species showed various reproductive success strategies in males and females, and that female promiscuity could have benefits other than reproduction.[79] These corrected assumptions improved scientific knowledge about primate breeding systems and sexual selection. But it also helped undermine some of the social assumptions regarding gender that science was uncritically promoting.

As these arguments show, meeting obligations to seek and engage dissenting views is crucial for advancing both epistemic and social aims. Yet determining what exactly these obligations entail is not always easy. Do researchers have duties to create opportunities for climate change skeptics to critically evaluate their work and take those criticisms seriously? For how long should they do so? Do they require the inclusion of dissenters on conference programs? Do these obligations demand that journal editors make an effort to incorporate intelligent-design scientists, for instance, as reviewers for manuscripts related to evolutionary theory? Do they entail that scientists must *always* involve dissenters in evaluating research, accepting hypotheses, or in synthesizing the current state of scientific evidence for policymakers?

What these obligations involve is significant, and not simply because such understanding is necessary to guide action. In a context where dissenting views can result in serious epistemological and social problems, determining the nature and scope of these obligations is particularly important. We will return to these issues in chapter 6, when we will consider whether one could use the reliable identification of NID to limit such obligations.

Conclusion

Scientific dissent, we have shown here, provides vital epistemic and social benefits. Dissent can further scientific progress by correcting false empirical assumptions, providing alternative ways of conceiving phenomena, and challenging value judgments. It can strengthen the justification for consensus views and thus increase confidence in their correctness. Additionally, the existence of dissent, even when it is incorrect, can foster public trust in science. The fact that dissent can play these important roles thus gives rise to obligations within scientific communities to seek and engage dissenting views. What exactly these obligations entail is contested. Yet, however one interprets these obligations, fulfilling them in a context where dissent can have adverse consequences creates a challenge for scientific communities in particular and societies in general: how do we protect and preserve the epistemic and social benefits of dissent while avoiding or limiting some of these problems? Might it be the case that some dissent, in virtue of the very sorts of features it has, will fail to yield any of the benefits that make dissent valuable? If so, this counts as an important reason for developing an account of normatively inappropriate dissent. If it is possible to reliably identify some dissent as normatively inappropriate, then perhaps this would allow us to develop strategies to tackle such dissent so as to prevent or limit its negative epistemic and social consequences.

3

Bad-Faith Dissent

In the 1960s, the tobacco industry launched a focused and coordinated campaign to generate data aimed at showing that smoking did not pose serious health risks to humans.[1] Their goal was to generate doubt about the harms of tobacco and ward off regulatory policies. As one now-infamous 1969 memo stated: "Doubt is our product since it is the best means of competing with the body of fact that exists in the mind of the general public. It is also the means of establishing controversy."[2]

The strategies employed by the tobacco industry have been applied to many other cases. Indeed, a number of scientists involved in efforts by the tobacco industry to create doubts about the link between smoking and lung cancer were also engaged in other controversies—for example, the causal connection between industrial emissions and acid rain, or between chlorofluorocarbons (CFCs) and ozone depletion—that were similarly aimed at preventing regulations unfavorable to industry interests.[3] More broadly, various industries and think tanks have sponsored research directed at manufacturing doubt about environmental toxins.[4] Scientists funded by the chemical industry have published studies challenging the consensus and purporting to support "hormesis," the hypothesis that low doses of toxins and carcinogens actually have some beneficial health effects for humans.[5] Similarly, conservative and free-market think tanks and organizations have devoted millions of dollars to engender doubt and public and political confusion regarding the existence and impacts of anthropogenic climate change.[6] In 2007, the American Enterprise Institute (a think tank largely funded by Exxon Mobil and staffed by former members of the George W. Bush administration) sent a letter to scientists offering $10,000 for any research contradicting claims made in the Intergovernmental Panel on Climate Change's (IPCC) *Fourth Assessment Report*.[7]

These cases of dissent are often seen as created in "bad faith" or with malicious nonepistemic intentions. Bad-faith dissent is thus dissent motivated not by a desire to help advance scientific knowledge but, rather, by some other objectionable goal: to confuse the public, stall policies that the dissenters dislike, promote particular ideological views, safeguard profits, or, more commonly, a combination of these goals, since the pursuit and realization of one will usually be made easier by that of the others. Can bad-faith motives be a criterion to identify normatively inappropriate dissent (NID)? Intuitively, research that aims to confuse the public or manipulate policy decisions seems unlikely to provide the sort of epistemic benefits for which dissent is valuable. In this chapter we consider this possibility. We explore various ways to explain why bad-faith motives could result in dissent that fails to promote or that impedes scientific progress, and we assess their plausibility. We consider whether the problem with these bad-faith motives is that they introduce nonepistemic values in the conduct and appraisal of science and we reject this option. We then contemplate whether what constitutes the problem is that bad-faith motives involve values that are not just nonepistemic but also antithetical to the epistemic aims of science. We conclude that in spite of the intuitive appeal of attending to motivations, they cannot serve as a criterion to reliably identify NID.

On Bad-Faith Motives

Cases such as the tobacco industry's attempt to produce dissent in order to confuse the public and stall regulatory policy that could be damaging to their financial profits seem paradigmatic instances of bad-faith dissent. As mentioned, the goal—recognized by the dissenters themselves—was not to promote scientific progress but, instead, to create doubt. Generating doubt helped advance their interests by undermining public confidence in a scientific consensus, a consensus used to ground public policies that set various restrictions on tobacco products.[8] This strategy can thus be useful to undermine the bases for health advice, energy policies, and regulatory decisions.

One possible way to explain why bad-faith motives would give rise to normatively inappropriate dissent (NID) is that motivations such as the desire to produce confusion, stall public policies, or increase profits are nonepistemic. This explanation is consistent with the traditional view that science ought to be value free.[9] Importantly, the value-free ideal does not maintain that science must be free of *all* values. There is widespread agreement that *some* values, such as empirical adequacy, internal coherence, and explanatory power, are actually constitutive of or instrumental to achieving scientific knowledge. Scholars consider these to be epistemic values.[10] Yet advocates of the value-free ideal contend that other values—nonepistemic ones, such as political, economic, moral, religious, or other personal values—fail to promote and may even hinder the production of knowledge.[11] Of particular concern is the potential that nonepistemic values have for "wishful thinking".[12] The worry is that if such values are allowed to determine researchers' decisions about what evidence there is, or which theories to accept, theories will be adopted because scientists *wish* them to be true, rather than on the basis of the evidence. Thus, those who agree with the value-free ideal could think that dissent motivated by nonepistemic aims is normatively inappropriate insofar as it allows nonepistemic values to direct scientific reasoning.

But the belief that nonepistemic motivations necessarily result in NID is mistaken. As many have argued, all science, or nearly all science, is in fact unavoidably motivated by some nonepistemic aims.[13] Some areas of scientific research are clearly aimed not only at arriving at true theories about the world but also at addressing certain social, ethical, or policy-related aims. Biomedical research's goal, for example, is not simply to discover truths about the human body or chemical compounds but also to improve public health and the quality of people's lives, which are eminently social and ethical aims. Nonetheless, this does not mean that the resulting research is necessarily biased or false. Similarly, individual scientists may be motivated by a desire to publish, obtain professional acclaim, or advance their careers, and yet their research may be, and often is, epistemically sound.

Even if it were possible to do research from epistemic motives alone, this would not be desirable. The presence of nonepistemic motivations in research can be epistemically beneficial.[14] For example, competitive motives to produce original research can help scientists solve problems faster. The desire for profits can drive industry researchers to ensure that their products are genuinely safe and effective, and thus that the research is epistemically justified. Moreover, creating financial incentives may help generate research in areas that address vital human needs—research that might otherwise not be pursued.[15] Producing research that is not only epistemically adequate but also successfully addresses important social needs is valuable.[16] Thus, research motivated by nonepistemic aims can be socially beneficial and epistemically sound. If this is the case, then the mere fact that some dissent is motivated by nonepistemic aims or values is insufficient for it to be normatively inappropriate.

Nonetheless, it might be that research that is epistemically "sullied" by nonepistemic motives can only play a crucial role in the advancement of knowledge if it is *also* accompanied by epistemic motives.[17] That is, insofar as research is motivated *only* by nonepistemic aims, it is unlikely to promote scientific progress. If researchers fail to be motivated by at least some desire to arrive at true or empirically adequate beliefs, they are likely to employ questionable methodologies or disregard evidence in order to arrive at whatever conclusions serve their nonepistemic interests. The problem with some dissent, then—such as in the case of the tobacco industry—is not the presence of nonepistemic motives but, rather, the absence of any epistemic ones.

Alternatively, one might argue that some nonepistemic motives may hinder or even be incompatible with the epistemic goals of science. This seems correct. At least some motives, such as the intention to deceive or cause confusion just for their own sake, are not problematic merely because they are nonepistemic; rather, it is because they are antithetical to achieving knowledge. Similarly, researchers who deliberately attempt to fabricate evidence are not simply motivated by nonepistemic values but also by values that are explicitly contrary to the epistemic aims of science. Dissent guided by these sorts of motives is likely to hinder knowledge production or fail to be epistemically useful.

Focusing on intentions that are indifferent or contrary to epistemic aims avoids the initial objections regarding the relevance of nonepistemic values in science. Nonetheless, using such bad-faith motives as a criterion to reliably identify NID is still problematic, for several reasons. First, determining what someone's motivations are or whether one is motivated solely by nonepistemic goals or not is a notoriously difficult task. People lack direct access to others' mental contents and thus they cannot reliably determine others' intentions. Indeed, individuals often fail to identify even some personal intentions correctly, either because they are unaware of or are mistaken about the motivations for their own actions. Not uncommonly, people deceive themselves as to their true motives, particularly when those motives might be vulnerable to criticism. Moreover, those who intend to deceive or have other problematic motivations are usually guarded about such motivations and have a stake in keeping their ill intentions hidden.[18] Granted, in some cases, such as the tobacco one, intentions may be obvious and well documented. But cases such as this seem infrequent and, like in the tobacco case, motivations become public only long after they have had considerable negative impacts.

Furthermore, it is difficult to determine whose intentions must be taken into account when assessing the normative appropriateness of dissent. Science is a social enterprise. Many researchers are involved in producing research that might turn out to conflict with widely held scientific claims and yet their intentions may be varied. Additionally, the research in question can be used by others, scientists and nonscientists. Whose intentions should be relevant in evaluating dissent when different actors are driven by different and conflicting motives?

Yet, even if it is not possible to determine people's intentions in dependable ways, perhaps there are some reliable indicators of bad-faith dissent.[19] After all, we often make judgments about people's motives on the basis of their behavior and these judgments usually turn out to be correct, even though we cannot directly assess anyone's mental state. Some have pointed out certain features of dissent that might help people in detecting the presence of what we are referring to as "bad-faith" motives,[20] such as the presence of financial conflicts of interest, directing the dissent to policymakers or the public instead of to the scientific community, or providing dissenting views without offering any real positive alternatives.[21]

There is a significant amount of evidence that financial conflicts of interest can result in research that is biased, fraudulent, and unreliable.[22] A variety of studies have found that biomedical research funded by industry is significantly more likely to report positive outcomes than studies funded by not-for-profit organizations.[23] Studies have also found that industry-supported reviews of drugs tend to be less transparent, less attentive to methodological limitations of the included trials, and present more favorable conclusions than the corresponding independent Cochrane reviews.[24] Furthermore, in the absence of statistically significant primary outcomes, industry trials are more likely to look for and report positive subgroup findings than nonindustry funded trials.[25] Financial interests can inappropriately influence the selection of evidence, interpretation or reporting of results, methodological choices, or the use of statistical methods.[26] Biased research fails to contribute to scientific progress and might actually hinder it. Perhaps, then, the existence of financial conflicts is useful as a proxy for identifying when dissenters are likely to offer dissent in bad faith.

Undeniably, financial conflicts of interest can problematically affect the epistemic quality of scientific research. Nonetheless, many acknowledge that the mere existence of such conflicts need not result in biased or otherwise epistemically invalid research.[27] Researchers who have financial interests can also be motivated by appropriate epistemic intentions. After all, the best way to ensure that particular interventions produce financial profits—and protect the reputation of researchers and industry—is to produce interventions that are safe and effective, which would usually call for conducting unbiased research. Indeed, valuable and important scientific knowledge has been produced in cases where financial conflicts of interest existed. For instance, the pharmaceutical industry's role was crucial for the development of statins, a class of drugs that has led to a significant reduction in cardiovascular disease morbidity and mortality.[28]

Moreover, research that involves financial conflicts of interest can also support consensus views. In fact, some dissenters have called attention to the existence of such conflicts when criticizing the consensus view regarding vaccine safety, for instance. Paul Offit—one of the main critics of vaccine dissenters—has been condemned for his ties to the vaccine industry, earning him the nickname Paul "For-Profit" Offit.[29] Those questioning the consensus on the safety of genetically modified products similarly point to the cozy relationships between agribusinesses such as Monsanto and the consensus view.[30] Climate skeptics have made similar claims about climate scientists, pointing to the fact that, for instance, Rajendra Pachauri, the chair of the IPCC from 2002 to 2015, was also the director of a for-profit alternative energy company.[31] If the existence of financial conflicts of interest is used as an indicator for bad-faith motives, and thus as an indicator that such dissent is suspect, then consistency would require that we make the same judgment when such conflicts are present in research that supports a consensus view. Given the economic context under which science is increasingly being produced today, this criterion would identify a significant amount of research as normatively inappropriate.

Of course, the presence of financial conflicts of interest can be a reason to be more attentive to the research in question, regardless of whether such research supports a consensus or a dissenting view.[32] Legitimate concerns arise when financial conflicts influence scientific research, and thus it is important to be attentive to the possible negative epistemic effects they might have. Nevertheless, financial conflicts of interest are an unreliable indicator of bad-faith dissent, as such conflicts can affect both biased and epistemically valuable research and both dissenting and consensus views.

The target audience for dissenting views is another feature of problematic dissent that could indicate the presence of bad-faith motives.[33] Presumably, dissent that intends to promote knowledge should be directed primarily at the scientific community and thus should appear in appropriate venues such as peer-reviewed science journals and scientific conferences. Some dissenters, however, target mainly nonscientific audiences.[34] Much climate change dissent, for example, is published not in academic journals but in popular books, newspapers, newsletters, pamphlets, and on blogs.[35] Organizations and think tanks use websites and hold talks at luncheons for public officials, most of whom are not scientists. As Naomi Oreskes points out, contrarians about climate change are "simply attacking the work of others and most doing so in the court of public opinion and in the mass media rather than the halls of science. . . . Contrarian views have been published in books issued by politically motivated think-tanks and widely spread across the Internet, but so have views promoting the reality of UFOs or the claim that Lee Harvey Oswald was an agent of the Soviet Union."[36] Similarly, dissent about vaccines has appeared mostly in popular books, on Internet sites, at public rallies, and on television talk shows, and seems geared toward the public.[37] Although, of course, the public must be adequately informed so as to engage appropriately in science policy and knowledge production, directing scientific claims mainly toward the public might indicate an intention to deceive or confuse rather than to contribute to genuine scientific debate, as the public usually lacks the training necessary to assess complex scientific claims.[38]

Yet, there can be legitimate reasons for using public venues and directing dissenting views toward a lay audience. When a strong scientific consensus exists about a particular theory or hypothesis, it can be very difficult for the scientific community to attend to dissenting views. For example, the peer review process for scientific journals generally calls for experts in the field to review relevant articles. This also means that such reviewers are more likely to hold the consensus position and be resistant to dissenting views that challenge such position. There is empirical evidence that reviewers, editors, and granting agencies have a bias toward epistemological conservativism—that is, they are less likely to evaluate positively and to publish views that challenge an existing paradigm.[39] Researchers proposing unorthodox claims or theories must generally meet a higher burden of proof than those supporting the standard view.[40] Similarly, confirmation bias can lead reviewers and editors to favor findings that agree with their own views and discount those that challenge them.[41] Climate change skeptics, for instance, often charge that they have been excluded from conference programs and advisory panels and that reviewers have treated their work unfairly in the peer review process.[42] Likewise, creationists and intelligent-design theorists argue that evolutionary theorists unreasonably dismiss their arguments against evolutionary theory.[43] Dissenters often believe—perhaps not always unreasonably—that the greater scientific community will not be sympathetic or responsive to their claims, and thus they decide to try their claims in the public sphere. Directing dissent toward the public may be an indication of this concern, rather than an intention to deceive or confuse.

Dissenters might also attempt to engage the public and policymakers precisely as means to ensure the attention of the scientific community, rather than with an intention to deceive or confuse the public. This occurs when the dissent in question comes from groups whose epistemic authority is not acknowledged by the scientific community, often because such communities might lack the kind of expertise many researchers believe is required for a meaningful scientific debate. For instance, Nadasdy provides a detailed analysis of the case of the Ruby Range Sheep Steering Committee in Canada,[44] which was formed in 1995 and tasked with addressing concerns about declining numbers of Dall sheep in the southwestern Yukon territory. The consensus view among biologists at the time was that the observed decline in the Dall sheep population was sudden and temporary, and that it was due to unusually harsh winters. Members of the Kluane First Nation, however, who had long relied on the sheep as a food source, rejected the consensus view as a satisfactory explanation of population decline. First Nation members on the committee argued that the sheep were well accustomed to harsh winters and that their numbers had already been in decline well before biologists began collecting data in 1974.[45] Unable to get traction with the scientific community, the Kluane First Nation took their concerns to policymakers, who then formed a commission comprising scientists, members of the tribe, and representatives of other interested parties, such as game hunters.[46] Their dissent eventually affected public policy and ultimately changed how scientists studied the sheep. In particular, it led to new methods for collecting data and calculating Dall sheep populations.[47] Thus, even though the dissent was largely directed toward policymakers rather than the scientific community, it was both socially and epistemically beneficial.

When public policies are at stake, involving the public in scientific debates is not just appropriate but also necessary, even if doing so can be difficult in practice.[48] If, say, dissenters believe that research supporting a consensus view is indeed questionable, and if the scientific community fails to attend to their concerns—or if dissenters perceive this to be the case—then it seems reasonable to bring the dissent to the public. Arguably, scientific evidence ought to inform public policy, and when dissenters call such evidence into question, this can affect the grounds for supporting certain policies. For example, in the debate over the acceptability of genetically modified products (GMPs), dissenting views often focus on what they perceive to be limitations of existing scientific evidence for GMP safety. Even though many scientists consider the safety of GMPs well established,[49] some call into question the strength of that evidence because they believe there are methodological concerns with many of those studies, such as short duration or the utilization of limited endpoints.[50] Yet, when it comes to developing public policy, it seems legitimate to ask whether the scientific evidence is "good enough" to reliably assess potential harms. Some people may reasonably be more risk adverse than others, or might balance risks and potential benefits differently, and it seems appropriate for these considerations to be publicly debated. Thus, the decision to take this debate to the public may be motivated by democratic values and concern for individuals' well-being, and not evidence of intent to confuse or deceive.

As with the case of financial conflicts, the target audience of dissent might lead one to be particularly attentive to the content of the dissent. However, given that there can be legitimate epistemic and social reasons to target the public rather than scientists, that dissent is primarily directed toward the public rather than the scientific community constitutes also an unreliable indicator of bad-faith dissent.

An absence of positive alternatives that can have some empirical success is another feature of problematic dissenting views[51] that could be useful in detecting the presence of bad-faith motives. Indeed, this has been one of the central criticisms of climate change skeptics.[52] According to this criticism, climate skeptics rely heavily on revealing and magnifying uncertainties in the existing data, rather than on providing new evidence and alternative theories.[53] Their time is thus devoted to calling into question the consensus view, rather than developing, promoting, and defending a competing positive account.

The failure of a plausible alternative may be reason for someone to persist in defending a consensus view despite the existence of dissent. But, it is not clear that it would be sufficient to identify such dissent as involving bad-faith motives and therefore as normatively inappropriate. Dissent that fails to offer an alternative need not be ill-intentioned and can be important to advancing scientific progress. Dissenters, for example, may correctly point out limitations in evidence or question the methodologies used to support a consensus view, even if they do not have a clearly developed alternative. In fact, such dissent might lead to additional research directed toward developing new or alternative methodologies. For example, a consensus among hydrogeologists exists about what models are currently most reliable in predicting long-term groundwater flow and radioactive waste migration in underground storage facilities (NRC 1995). But these models have significant flaws and limitations.[54] (It is reasonable to ask whether the best available models are good *enough* to generate predictions that will be used to ground decisions about how and where to store nuclear waste. Even if there is no better alternative model available, there may be cases where the best model fails to be good enough to rely on, particularly when there might be devastating and irreversible consequences for people's health or the environment.

This also seems to be the case with at least some climate change dissent. Their claim is that the current modeling methods involve significant uncertainties, and they take those uncertainties to be important for how high our confidence should be in relying on the resulting predictions.[55] Clearly, the degree of confidence we have in such predictions may well be relevant to decisions regarding what policies should be adopted or how to assign priorities in allocating funds for mitigation or adaptation.[56] While skeptics may ultimately be incorrect about the uncertainties that exist or their significance, dissent pointing to such limitations seems both relevant and important to making policy decisions, even when they offer no positive alternative for how to model the climate system.

Even if bad-faith motives could be accurately identified, it is not clear that this criterion can actually capture cases of dissent that many consider to be normatively inappropriate. For example, many accuse intelligent-design theorists of being motivated by a desire to protect certain religious doctrines, regardless of the evidence. This might be understood as an aim incompatible with epistemic goals. Yet, this seems an uncharitable interpretation of their intentions. Presumably, intelligent-design theorists are sincerely convinced that evolutionary theory is false and their religious beliefs are true (otherwise it is difficult to see why they would be motivated to protect them). Thus, they are motivated by a desire to produce and disseminate theories they take to be accurate about the creation of human beings. They can also believe that religious values must play a role in inquiry because they take contribution to religious knowledge or to salvation to be primary aims of inquiry. Similarly, some of these dissenters are mistrustful of scientists maintaining a consensus position because they see such consensus as advancing economic or political agendas of their own and they want to correct these biases. Thus, they have certain epistemic motivations. Insofar as creationism is thought to be a paradigm case of NID, then, it must be on the basis of something other than simply the motives of its proponents. Likewise, in the case of climate change skepticism—also assumed by many to be a case of inappropriate dissent—dissent can arise not because of malicious motives inconsistent with knowledge production but, rather, because of different background assumptions and values that give rise to genuine disagreements about what the evidence is and whether theories or models are justified. For instance, while many climate skeptics have economic or political interests they wish to advance, recent empirical studies in social psychology suggest that climate skepticism cannot be fully explained in terms of an intention to deceive or to promote political or economic interests.[57] Hoffman utilizes field observations from the largest conference of climate deniers in the United States, as well as 800 publications by climate skeptics, and concludes that much climate skepticism is the result of differences in the ways in which the different groups frame empirical and policy issues. Climate deniers and consensus scientists define the central research questions differently, conceptualize the problem of climate change dissimilarly, and make diverse value judgments about the acceptable range of solutions and their necessity.[58] Disagreements and dissent arise in part, he argues, because of these differences, rather than because of any malicious intent to deceive or confuse. If this is correct, bad-faith intentions are not necessary for NID.[59]

Finally, it is not obvious that the presence of bad-faith motives—whether or not they are accompanied by epistemic ones—is sufficient for NID. It is not clear why motives or intentions, both appropriate and inappropriate, are relevant to the epistemic value of dissent. Clearly, people's motives are distinct from the consequences that follow from people's actions. Although dissenters might intend to simply confuse the public or policymakers rather than to generate knowledge, that they intend such things does not entail that the dissent in question cannot contribute to scientific progress. Bad-faith motives were clearly present in the research that Wakefield conducted regarding the safety of the MMR vaccine.[60] Wakefield altered multiple facts about the patients' diagnoses and medical histories in order to support his fraudulent claim that the MMR vaccine caused autism, and he sought to exploit the resulting MMR scare for financial gain. In spite of this, the dissenting view about the safety of vaccines, fueled at least in part by Wakefield's research, has led to a strengthening of the evidence about vaccine safety.[61] Of course, this does not justify Wakefield's behavior or his ill intentions. But it does show that, even when bad-faith motives are present, they need not result in dissent that fails to promote or hinders scientific progress.

Granted, dissent exclusively motivated by a desire to confuse or deceive is likely to be biased, as the motivation is to generate misunderstandings. Nonetheless, the fact that the dissent is the result of biased research would give us grounds to question the dissent independently of what motivated it. That is, whether the dissent was valuable or not would depend on the evidence for the dissent and not on the motives of the researchers or of those promoting dissenting views. Scientific claims must, after all, be publicly assessed.

Of course, that bad-faith motives can sometimes produce epistemically sound dissent does not transform those intentions into appropriate ones. Their presence when producing dissent tells us something about the moral character of the dissenters. Whether or not attempts to mislead or deceive are epistemically problematic, they are certainly ethically suspect. Codes of practice for most scientific disciplines, organizations, and professional societies include honesty as a professional virtue. Thus insofar as scientists attempt to deceive, they are failing to discharge their professional duties. Moreover, misleading people treats them merely as means to the deceivers' ends and disrespects individuals' autonomy and prevents them from making informed decisions about how to act or what policies to support. Nonetheless, although at least some motives are ethically improper, motives or intentions need not track consequences and hence dissent thus motivated could nonetheless help advance knowledge.

Conclusion

It seems clear that there is something problematic about dissent primarily motivated by a desire to deceive or confuse the public, stall regulatory policies, or promote ideological or economic interests rather than to advance knowledge. Such dissent can have negative epistemic and ethical consequences. Yet despite this, focusing on the intentions under which dissent is produced is ultimately unhelpful in reliably identifying NID. Bad-faith motives are neither necessary nor sufficient for such dissent. At least some cases many see as involving NID seem to lack bad-faith motives, and dissenters can produce epistemically valuable dissent even when they have problematic motivations. Moreover, determining what motivations dissenters might have is not an easy task. Motives are difficult to ascertain, whether on their own or by relying on particular indicators.

Focusing on bad-faith motives as a criterion to identify NID might not just be useless but also dangerous. Placing too much weight on the role of intentions can lead one to believe that the absence of pernicious intentions is sufficient for accepting certain dissenting views as epistemically valuable.[62] After all, it seems that dissent can be normatively inappropriate and result in negative consequences, such as confusion or stalled policies, even when dissenters have appropriate motives. As we have seen, several of the cases that many take to be paradigm cases of NID arguably do not involve bad-faith intentions. Thus, if they are indeed normatively inappropriate, it must be on alternative grounds.

4

Failing to Play by the Rules

Imagine two chess players trying to win a game. For the game to be minimally successful, opponents must have the pieces arranged appropriately, must know how to move each type of piece, when to move it, and how to checkmate. The opponents must play by the rules of the game in order to achieve the objective of the game, which is to checkmate the opponent's king. Engaging in scientific practice also requires that relevant parties play by certain rules in order to achieve the objectives of science, which is to produce significant knowledge. Perhaps, then, rather than attending to motives, a more successful strategy to identify normatively inappropriate dissent (NID) might be to focus on whether scientists actually follow appropriate rules of engagement for fruitful discussions concerning competing scientific views. Some have referred to this as "agonistic dissent," or dissent that plays by the "rules of the game" within a given scientific field.[1]

In chapter 2, we saw that dissent is epistemically beneficial at least in part because it plays an important role in helping identify and critically evaluate implicit background assumptions, proposing alternative theories and explanations, and promoting the revision of accepted theories so that they are even more justified. That is, dissent contributes to scientific progress. Yet, in order for dissent to yield epistemic benefits, it must be governed by rules or norms for what constitutes legitimate challenges and appropriate responses. Dissent that fails to respect basic epistemic norms would be normatively inappropriate, as such dissent would fail to yield the benefits of transformative or effective criticism and would thus not contribute to knowledge production. To put this in terms of the chess analogy, it is not that some dissent constitutes a poor or unstrategic move but, rather, that it constitutes a violation of the rules of the game such that it ceases to constitute a chess move at all.

Philosophers of science have proposed three criteria as necessary for transformative criticism. According to these criteria, participants must: share some standards of evaluation;[2] engage in uptake of criticism against their own views;[3] and possess some reasonable degree of expertise in the subject related to the dissent.[4] First, to have a fruitful discussion about competing views, some shared standards for assessing evidence must exist. That is, there must be shared norms that guide the appraisal of scientific claims in order for dissenters[5] to be able to engage in fruitful discussion with those who hold the consensus view. Second, inquirers must take challenges to their own views seriously and respond to them by revising their arguments or explaining why the criticism is unsound or misguided.[6] If dissenters merely repeat the same arguments, without acknowledging and addressing criticisms raised, the discussion cannot be advanced and thus dissent fails to further knowledge and might even hinder it. Third, in order to contribute to productive scientific discussions, participants must have some degree of expertise in the area of research under discussion. There is a difference between an informed dissenting view and a refusal to believe a scientific claim simply in virtue of not understanding the evidence or the methodology at stake. To take dissenters' criticisms as instances of genuine scientific dissent or, as many have put it, for others to grant dissenters equality of intellectual authority, dissenters must have some degree of expertise about the subject of inquiry.[7] As we have argued in chapter 2, it would be problematic to dismiss dissent on the basis of some epistemically irrelevant feature, such as the gender, race, or political affiliation of the dissenter.[8] Nonetheless, being an "eligible player" in the scientific inquiry game requires that one meets certain qualifications relevant to the area of science under investigation.

These three criteria are interrelated and mutually supporting in bringing about effective criticism. For example, sharing standards requires knowledge of basic methodological approaches in a particular area of research. That is, it may require some degree of appropriate expertise. Similarly, as we explain below, responding to criticism—or engaging in uptake—involves appealing to shared standards of what counts as a criticism or as a reason for accepting or rejecting a hypothesis. Nonetheless, we will examine each of these criteria individually, as they highlight different features of fruitful criticism that, if lacking, would give rise to dissenting views unable to contribute to scientific progress.

Many have used one or more of these criteria to reject certain cases of dissent as normatively inappropriate. For instance, some have argued that the dissenting views against the evolution of species expressed by creationists or intelligent-design theorists should be dismissed because they fail to use shared standards or to engage in appropriate uptake to criticisms.[9] Many have also argued that climate change dissenters similarly fail to appeal to share standards in making criticisms of consensus views[10] or that they lack relevant expertise related to climate change science.[11]

Yet, although these criteria appear eminently reasonable as requirements of transformative criticism, and thus of appropriate dissent, what they involve is not straightforward. In this chapter we evaluate them in detail and argue that they fail as criteria to reliably identify NID. Indeed, on some of the interpretations of these criteria, they cannot be used to pinpoint the very cases of dissent that are taken by their advocates to be paradigm cases of NID, such as dissent with respect to evolutionary theory, climate change, and the safety of genetically modified products (GMPs).[12] We first show that judgments about whether dissenters have engaged in uptake can be made only relative to a set of shared standards of evaluation. Thus, the usefulness of uptake as a criterion to identify NID will depend on the success of the shared standards criterion. We then contend that there are several different, and inconsistent, ways of understanding what constitutes "sharing standards" and that these differences make this criterion unhelpful in trying to reliably identify NID. Likewise, the requirement of expertise can be variously understood, and thus it presents us with similar problems.

Uptake of Criticism

The uptake criterion requires that all participants address criticisms and responses raised by others.[13] Those who maintain a consensus view must be responsive to the criticism of dissenters, but dissenters must also engage with the criticisms or replies they receive against their views. This does not entail that people must submit to a particular objection but, rather, that they engage with the criticisms raised by other participants. They can do so by defending their methodologies, interpretation of the evidence, and background assumptions and by giving reasons for why the objections are unfounded or by revising their views accordingly.[14] Dissenters also engage in uptake when they come to agree with a particular response to their criticism, but present new evidence for which the majority view cannot yet account. Clearly, failing to engage in uptake prevents the sort of critical discourse necessary to advance a debate. Insofar as dissenting views do not engage in uptake, then it would seem reasonable to regard such views as NID.

But how are we to determine whether a failure to engage in uptake has actually occurred? It cannot be the case that engaging in uptake requires one to capitulate to any criticism raised, as this would demand that all participants simply agree with each other whenever objections are proposed, which is not warranted in many cases. Similarly, it cannot be that uptake involves merely the acknowledgment of a challenge, as this would also fail to contribute to scientific progress. Uptake, then, requires that participants actually *engage* in some meaningful way with the arguments presented and that they critically evaluate the extent to which such arguments are correct or not and proceed accordingly.

Determining whether someone has or has not partaken in this kind of meaningful engagement is, however, not an easy task. Consider, for instance creationist and intelligent-design (ID) theorists who criticize evolutionary theory. They are often taken to be a paradigmatic case of a failure of uptake.[15] Many have argued that evolutionary biologists have already given a great deal of attention to the views of creationists and, recently, of more sophisticated ID theorists. Contemporary scientists, historians, and philosophers have dedicated considerable time and effort to addressing criticisms of evolutionary theory and other relevant theories that operate as background assumptions for supporting evidence, such as the physics of radioactive decay.[16] But they contend that these efforts have largely been ignored and that critics of evolutionary theory spend their time merely repeating the same claims over and over again.[17] For example, creationists have often argued that evolutionary theory is inconsistent with the second law of thermodynamics. Although this has been refuted, dissenters appear to repeat the criticism without even acknowledging the arguments that evolutionary theorists have made.[18]

Yet despite this characterization, many intelligent-design theorists believe they are engaging in uptake by responding to the challenges raised by evolutionary theorists. In the dissenters' view, the reason they still advance some of the same criticisms is that they believe they have appropriately explained why the evolutionary response is unsatisfying.[19] Indeed, some ID theorists have claimed that it is evolutionary theorists who have failed to engage in uptake of the criticisms presented to their views.[20]

Or take, for instance, dissenting views about the safety of childhood vaccines. Proponents of the consensus insist they have provided compelling evidence about the safety of such vaccines.[21] They use epidemiological data to buttress their case and complain that dissenters are simply ignoring the evidence and repeating old arguments about the relationship between autism and the use of certain vaccines. Some dissenters, however, contend that consensus proponents are not attending to their objections.[22] They argue that using epidemiological evidence as proof of the safety of vaccines misses the point. Dissenters can agree that vaccines are a good public health measure—that is, that they are safe relative to the population and the benefits obtained. But they insist that claims about the safety of vaccines ignore the fact that some children are, in fact, harmed by them. What some dissenters want is research on why particular children are indeed harmed. That is, dissenters are concerned with the safety of the vaccines for *their* children and thus they do not take epidemiological research as a response to their concerns.[23]

It seems, then, that judgments about whether one has actually engaged with the arguments presented presuppose some shared standards of evaluation. That is, there must be some agreement among contenders about what constitutes a criticism or response to a scientific claim, when a particular criticism is relevant, and what counts as a resolution or a destabilization of a particular position. This might include, for example, shared assumptions about what constitutes good evidence, norms guiding research practice, shared methodologies, criteria for what a successful theory needs to explain, and whether inconsistencies with other widely accepted scientific theories are problematic. Without such shared standards there is no basis for determining what counts as a challenge or a response and, thus, whether uptake has occurred or not.

If this is the case, then the notion of uptake, by itself, is not sufficient to identify when dissent is normatively inappropriate. The criterion of shared standards is also needed in order to recognize whether uptake has taken place. Of course, the criterion of uptake might still be necessary to determine whether a particular instance of dissent is or is not normatively appropriate. If participants share standards but ignore the objections raised by others, transformative criticism cannot take place. Nonetheless, determinations about uptake cannot be performed without understanding the criterion of shared standards.

Shared Standards

A second possible requirement for "playing by the rules" is that criticism must conform to some shared standards of evaluation.[24] In order for dissent to advance knowledge in a particular research context, it must appeal to some norms or values accepted also by those who hold the position being criticized. These shared standards include both the aims of inquiry and criteria for choosing between competing theories. Such standards may consist of commitments to epistemic values, or standards that are instrumentally valuable for producing true theories such as empirical adequacy, internal coherence, or external consistency with other accepted theories.[25] They may include what some have called "cognitive values"—that is, values that promote other cognitive or pragmatic aims of science, such as expansion of knowledge, simplicity, or explanatory power.[26] They may also involve certain social values that are connected to the aims of some research programs, such as relevance to or satisfaction of particular social needs,[27] or usefulness in informing public policy.[28] The particular standards endorsed, as well as how those standards might be weighed against each other, may vary depending on the research context. However, insofar as dissenters and consensus proponents fail to have any shared standards, no productive, transformative criticism is possible.[29]

Various authors have argued that creationist dissent can be excluded as normatively inappropriate on this basis.[30] According to this objection, such dissent relies on standards that the scientific community does not share, such as conformity to religious doctrine, and rejects others that scientists widely accept, such as empirical adequacy, or consistency with other highly confirmed scientific theories. This can also help explain why, for example, it appears that creationists have failed to engage in uptake. That is, they hold different standards for what constitutes an objection to evolutionary theory, or for what constitutes a satisfactory response. To the extent that creationists lack standards widely shared by the rest of the scientific community, their dissent cannot contribute to advancing knowledge and thus it should count as normatively inappropriate.

Nonetheless, if we are to use this as a criterion for identifying NID, it is important to be clear about what this criterion requires. That is, when should we say that dissenters sufficiently share standards of evaluation? Addressing this question requires clarification regarding (1) the number of standards that need to be shared for dissent to count as normatively appropriate or inappropriate, and (2) the degree of agreement about the content of the standards shared.

Presumably, there must be at least one shared standard. Indeed, some have argued that empirical adequacy alone is sufficient to identify whether dissent is normatively appropriate or not.[31] But empirical adequacy is not an unambiguous concept. Indeed, both intelligent-design theorists and climate change skeptics have expressed a commitment to empirical adequacy. Of course, it is tempting to argue that ID theorists' claims to empirical adequacy are not reasonable. For example, ID theorists sometimes make ad hoc modifications to background assumptions in order to render observed data consistent with their views. But, in this case, what will count as a "reasonable" claim to empirical adequacy will need to appeal to other shared standards, or at the very least to other shared assumptions about what constitutes empirical adequacy.

If other standards need to be shared, then how many and which ones? There might be reasonable disagreements about whether simpler or more complex theories or models should be preferred in a context where participants all agree that any theory or model presented is empirically adequate.[32] If as Mitchell has argued,[33] many biological systems involve both constitutive complexity (in the sense of comprising many parts) and functional or organizational complexity, then a preference for simpler, reductionist accounts in biology might be unjustified. In some cases, it may be that simpler models or theories limit our ontologies and obscure our understanding of phenomena, even if they are otherwise empirically adequate.[34] Simpler theories, however, might be preferable to more complex ones that are also empirically adequate when the simplicity allows for a better predictive value.

Moreover, while disagreements about standards of evaluation can be a source of conflict, they can also have substantive epistemic benefits, as such disagreements allow participants to compare and critically assess putative knowledge created under different standards.[35] It can also enable participants to see things from multiple points of view. Thus, some dissent about standards of evidence is arguably appropriate and can benefit knowledge production. Presumably, then, the sharing of some standards but not others might provide epistemic benefit, but it is difficult to determine which standards and how many should be shared for a particular dissenting view to count as normatively appropriate or not.

But there is an additional difficulty in relation to the criterion of shared standards: it is not clear what the depth of the agreement about the content of a standard must be. Scientists must interpret particular standards or values for theory choice in order to apply them, and such interpretations may vary.[36] For example, empirical adequacy can come in varieties involving depth (fidelity to a body of particular evidence) or breath (ability to apply to a range of domains).[37] Therefore, even if contenders share empirical adequacy as a standard, they might disagree about the relevant sense. Similarly, even among those who agree that simplicity is a virtue in theory choice, there are disagreements about what "simplicity" means.[38] Does having shared standards require not only the agreement that simpler theories are to be preferred but also agreement about how to interpret simplicity? Such a requirement could identify as normatively inappropriate some of the cases of dissent that are intuitively problematic. For instance, while evolutionary theorists and intelligent-design theorists may both profess a preference for simpler theories, they seem to have different understandings about what this means. Some ID theorists, for example, adhere to an understanding of simplicity as quantitative parsimony when they argue that their theory is simpler than evolutionary theory. They claim that ID is quantitatively simpler in the sense that it posits a *single cause* that can explain observed complexity in the world.[39] However, evolutionary theorists claim that their theory is simpler because it posits *fewer types* of ontological entities. That is, it is more qualitatively parsimonious because it does not posit a supernatural being necessary to explain certain phenomena.[40] Thus, we might be able to conclude that the dissenting views of ID theorists are normatively inappropriate because, even when they may share broad theoretical commitments, they have no shared understanding about what simplicity means.

But, of course, if communities had to share not only broad theoretical commitments but also particular interpretations of those commitments, this would reduce the sort of epistemic pluralism within scientific communities that is valuable.[41] On this interpretation, the shared standard requirement would be "thicker," in that it would require that contenders not just share particular standards but also reach greater agreement about how to interpret the standards in question.

Alternatively, the requirement for shared standards could be interpreted in even "thicker" ways, as one that involves sharing not only the interpretation of the standards but also enough other background assumptions to be able to apply the standards in question in particular cases. This way of understanding the shared standards criterion would again allow us to identify as normatively inappropriate those cases of dissent that many have taken to be problematic. Consider, again, the dispute between creationists and evolutionary theorists.

All may agree that empirical adequacy is a virtue and may even agree that to be empirically adequate a theory must account for observed phenomena. But one might argue that creationists fail to have a shared understanding about what is actually observed and whether or not it is explained by the theory in question. For instance, creationists argue that observations of the fossil record reveal that there are no transitional fossils, and they take this to be evidence for special creation. Evolutionary theorists disagree about what we would expect to observe if the theory of evolution were true. First, they point out that evolution occurred slowly over thousands of years, such that we would not expect to see sudden shifts of the sort of transitional fossils that creations claim we ought to see if evolution were true. Moreover, they argue that there are significant gaps in the fossil record and that it represents only a very limited view of the sort of life that existed. Thus, to the extent that we fail to observe transitional fossils, this is easily explained by evolutionary theory. Hence, creationists fail to sufficiently share the sorts of background assumptions that are necessary for being able to apply empirical adequacy in the same way as do evolutionary theorists.

On this "thicker" interpretation, climate change skeptics also would lack shared standards. For example, although climate change skeptics share a commitment to empirical adequacy, and may agree with what that involves, they disagree about what the empirical evidence is because they take some data to be unreliable. They have criticized data-gathering techniques that were used to establish the temperature record on which climate scientists often rely.[42] Thus, on a thick interpretation of shared standards, we might say that the dissent of climate skeptics is normatively inappropriate because it does not rely on shared standards regarding what constitutes appropriate data collection or statistical methodologies. Indeed, climate change skeptics have been criticized for using problematic statistical methods.[43]

But, of course, this understanding of the shared standards criterion will also capture cases of dissent that many take to be normatively appropriate, like the dissent regarding GMPs. Indeed, Philip Kitcher—who believes a consensus on the safety of GMPs exist[44]—seems to be using precisely this understanding of sharing standards when he argues that, in the ideal deliberation that occurs in well-ordered science, there would be no rational dissent against the consensus on the safety of GMPs. For Kitcher, those who press dissenting views about the safety of GMPs do so not because they fail to recognize the virtue of empirical adequacy, or have different views about what empirical adequacy requires. Rather, Kitcher argues, the problem with these dissenters is that they have other problematic background assumptions that are leading them to incorrectly reject the empirical evidence that GMPs are actually safe. He claims that such dissenters must either have the false belief that GMPs are somehow different in kind from other biological organisms or that the empirical evidence is not reliable because it has been produced by those with financial conflicts of interest. But in well-ordered science, where participants are fully informed and there are checks on commercial interests, Kitcher argues that neither of these background beliefs could be reasonably held.

This suggests that, in order for dissent to be normatively appropriate, dissenters must share not only a commitment to empirical adequacy but also background assumptions that are relevant to determining whether or not certain scientists should take particular evidence as germane to a particular hypothesis, such as the safety of GMPs. But dissenters have different assumptions about what constitutes a risk, how risk is best measured, and the extent to which those making risk assessments ought to be trusted as reliable.[45] Kitcher seems to think that such disagreement could not reasonably occur in well-ordered science and thus that the dissent produced by opponents of GMP safety would fail to promote scientific progress.

This thicker interpretation of the criterion of shared standards might indeed aid us in identifying cases of dissent that seem intuitively problematic, such as ID. However, as the case of GMP dissent shows, it might identify as normatively inappropriate cases of beneficial dissent where there is disagreement about background beliefs that are crucial to measuring empirical adequacy. Consider, for instance, the case already discussed regarding the debate in anthropology over the evolution of tool use in humans.[46] As we saw in chapter 2, the consensus view was that tool use evolved as a result of the hunting practices of males, who needed to find ways to secure food for their families.[47] But when more women began to enter the field of anthropology in the 1970s and 1980s, they began to challenge this view and to argue that the activities of women also contributed to the evolution of tool use in humans.[48] Advocates of both hypotheses shared a commitment to empirical success, as well as a commitment that empirically adequate theories should be able to help make successful empirical predictions. But, they disagreed about what the archeological record was showing. Proponents of the woman-the-gatherer hypotheses pointed to evidence regarding the provision of plant resources provided by women, and argued that reeds and baskets used for foraging and transporting food should count also as tools, and thus that, contrary to the man-the-hunter hypothesis, women had also been essential in the evolution of human tool use. Different background assumptions about what constituted a tool thus led the different parties to diverse conclusions about the archeological record. Hence, although the two groups shared a commitment to empirical adequacy, they failed to share background assumptions that were salient for determining what counted as the empirical data in need of explanation.

In addition, the thicker interpretation of shared standards seems to allow only insiders to bring forth criticism in scientific debates. That is, it must be those who are familiar with and share certain methodological approaches the ones who must provide the challenges, as noninsiders might be unfamiliar with the particular interpretations. But this seems to limit criticism that would potentially be epistemically valuable.

What the preceding discussion shows is that the criterion of shared standards is not as straightforward as it might first appear, and thus it is not particularly useful as criterion for reliably identifying NID.[49] Interpreted as a thick requirement to share the same theoretical virtues, the same interpretation of those virtues, and sufficiently similar background assumptions that would be relevant to applying those virtues, it would certainly identify as normatively inappropriate all the cases of dissent that many take be problematic— for example, intelligent design, climate change, vaccine rejection. But it would do so at the cost of limiting the diversity of perspectives that is so crucial to scrutinizing background assumptions and proposing alternatives; that is, it would constrain precisely the epistemic benefits that dissent offers. It would exclude as inappropriate dissent that actually contributes to scientific progress. Conversely, interpreted rather thinly as a commitment to having some broadly shared standards—to empirical adequacy at the very least, understood in some particular way—but allowing disagreements about how to best understand them or apply them in practice, the criterion would exclude cases that many consider paradigm cases of NID.

Expertise

Typically, rules for playing a game also include rules about *who* is eligible to be a player. Thus, agreeing that dissenters must be accorded equality of intellectual authority, or the authority to have their criticisms taken seriously, need not commit one to denying that differences in the cognitive authority of dissenting views can be relevant in according such authority. Some have insisted that dissenters must have adequate expertise relevant to the area of research under consideration for such dissent to be taken seriously as genuine scientific dissent.[50] Harker, for instance, argues that dissent is normatively inappropriate—or in his terms, inappropriately creates controversy—when "the overwhelming majority of people *qualified* to judge a scientific issue are in broad agreement over what conclusions are justified, but those conclusions are grossly misunderstood, denied, regarded with acute suspicion, and even ridiculed outside the community of experts."[51]

It seems uncontroversial that scientific experts should be granted authority in scientific disputes. Clearly, if one is making a claim about, for instance, the DNA sequence of a genetic mutation, one would need to have relevant knowledge in this regard. However, in many instances, determining what constitutes appropriate expertise, or who should count as a relevant expert, is not an easy or uncontroversial task.[52] A variety of empirical studies attempting to show the high degree of consensus in the scientific community about the existence of anthropogenic global warming take publications in peer-reviewed scientific journals in climate science as a proxy for relevant expertise.[53] Yet, these studies take different numbers of peer-reviewed publications as necessary to count as an expert. For example, some of the studies require ten or more peer-reviewed publications on climate change, while others included authors with at least one peer-reviewed publication.[54] Harker, on the other hand, calls for a track record of peer-reviewed publications, a Ph.D. in a relevant field, and a history of successful grant writing in that field in order to count as an expert.[55] Applying these understandings of expertise to climate change disputes has led critics to argue that most climate change dissent should not be viewed as genuine scientific dissent, because the dissenters in question fail to be experts in any climate science–related field.[56] For example, Oreskes and Conway dismiss the climate change skepticism of S. Fred Singer because, although he was a trained atmospheric physicist, his research focus was rocket science rather than climate modeling.[57]

But expertise so understood seems a problematic criterion to identify NID, for several reasons. First, it is not clear that it would exclude some of the cases that many take to be paradigm cases. For instance, at least some ID dissenters have pertinent disciplinary expertise in biology, biochemistry, geology, and so on, and have published in peer-reviewed, discipline-relevant, journals.[58]

Second, understanding expertise in this narrow way would render as normatively inappropriate cases of dissent that are epistemically valuable, despite originating from someone who was not an expert in the narrow sense.[59] Consider, for example, the Great Devonian Controversy, which concerned the mapping and interpretation of the geological strata of the county of Devonshire in 1834.[60] While doing work in Devon, the surveyor De la Beche concluded that a certain deposit—the Culm north of Dartmoor—that was thought to belong to the Silurian period, contained fossils that were characteristic of the younger carboniferous age. He concluded that using fossil information to date rocks was misguided. Roderick Murchison, an aspiring geologist who defended the fossil criterion to correlate strata, countered that De la Beche had wrongly placed the fossils. The result of this controversy was the establishment of a new geological period, the Devonian, and of the method of dating rocks through fossil evidence. De la Beche, a field worker, initially rejected Murchinson's claims, at least in part because Murchinson was a theorist who had not even set foot in Devon and therefore lacked the relevant expertise. Moreover, although those most intimately involved in the controversy were geologists, they were required to rely on the information provided by fossil experts in order to arrive at their conclusions.

Interpreting expertise too strictly is particularly problematic when dealing with certain scientific areas. For instance, climate science research involves work being done in a variety of disciplines and subdisciplines, which use different methodological approaches to examine diverse aspects and impacts of the climate system. In such context, it seems markedly challenging to identify the boundaries of who counts as a "climate change expert." Indeed, a significant amount of science being conducted today, from climate change to high-energy particle physics, to genomic science, requires the participation of large national and international teams of researchers with multiple forms of expertise, which ranges from statistical approaches, to instrumentation, to specific scientific knowledge in particular sciences. Hypotheses in these contexts are not generated and tested by scientists who all have the same scientific expertise but, rather, by a team of researchers, none of whose members individually might be able to actually fully understand the research results, but all of whom are intrinsic to the success of the research in question. But if scientists from a variety of disciplines and different types of scientific expertise are necessary to arrive at a consensus view, then scientists from a variety of disciplines and different kinds of expertise should, in principle, be granted equality of intellectual authority to present dissenting views.

Thus, constructing expertise in narrow disciplinary terms will again result in the elimination of normatively appropriate dissent. For example, although there are disagreements about whether a consensus view exists regarding the safety of GMPs,[61] molecular biologists largely agree that such consensus exists. In general, however, ecologists and agricultural scientists disagree that there is sufficient evidence to support safety claims. Using narrow disciplinary criteria of expertise would lead to discounting their views, which arguably would prevent legitimate dissenting views from being taken seriously.

Third, it is not clear that only formally trained scientists should be recognized as having the expertise or equality of intellectual authority that might bear on advancing scientific knowledge or on participating in valuable dissent. This is particularly the case when the scientific disputes have direct effects on policies or practices that affect multiple stakeholders. There is growing evidence of the ways in which other types of knowledge, based in experience rather than formal training, for instance, can produce valuable dissent.[62] For instance, in a well-known case discussed by Brain Wynne, he documents the disagreements between scientific experts and local sheep farmers over the effects of radiation from the Chernobyl accident on sheep in Cumbria, a mountain area of northwest England.[63] Shortly after the Chernobyl meltdown in the Soviet Union in April 1986, rain deposited radioactive material on portions of Great Britain. After finding high levels of radioactivity in samples of lamb meat from the Cumbrian fells, the British Ministry of Agriculture, Fisheries, and Food (MAFF) imposed a ban on the movement and slaughter of sheep in parts of Cumbria and North Wales. Various scientific assumptions about absorption and decay of radiocaesium led scientists to assure farmers that a three-week ban would be sufficient to solve the problem, but levels of radiation in the sheep unexpectedly continued to increase, and thus the ban was extended indefinitely. Scientific experts sent by MAFF to evaluate the levels of radioactivity levels relied on their own theoretical models for doing field research, and did not solicit any feedback from the sheep farmers. But the sheep farmers disagreed with many of the conclusions to which scientists arrived. Their dissent was, of course, not grounded on disagreements about the degree of absorption or decay of radioactive material but, rather, on their knowledge of relevant local conditions—for example, sheep grazing behavior, areas of water accumulation, soil composition, farming practices—knowledge that was actually essential to determining radiation activity and that scientists could have used to perform appropriate analysis. Scientists, however, ignored the farmers' expertise, which resulted in erroneous evaluations. As this example shows, those who may not meet criteria for formal training within a scientific discipline or who fail to have peer-reviewed articles can still contribute valuable dissent that advances scientific knowledge. Indeed, their dissent may be valuable precisely because they are outsiders.

Fourth, narrowly defining expertise in terms of formal scientific training neglects the fact that at least some scientific fields involve ethical and social values—that is, contextual value judgments.[64] Some scientific areas such as ecology and biomedicine, for instance, involve value-laden concepts—for example, ecological sustainability, or race.[65] Consider, for example, epidemiological research on racial health disparities. Various ways of representing "race" exist, such as the "one drop rule"—the biological race of the mother, self-identification, or geographical ancestry.[66] Which classification method is most appropriate will depend on what it is that one is interested in tracking—for example, ethnic or racial disparities in health risks that result from particular genetic variations, disparities that are the result of unjust discrimination, and so on—and such decisions involve contextual value judgments.[67] Methodological choices can also depend on the social aims of the research.[68] Similarly, decisions about whether one should prefer simpler models to complex ones also depends on what features one thinks are important to make visible or to be obscured, and thus require contextual value judgments.[69] Moreover, value judgments play a role in assessing the risks of error all throughout the research process, from decisions about methodologies to judgments regarding the characterization of data, to determinations about theory acceptance.[70] Some of the debate about climate models, for example, has centered on the role of value judgments in weighing the risks of errors related to uncertainties.[71] Persistent uncertainty exists about, for instance, the extent to which cloud formation and water vapor feedback will intensify or dampen warming trends,[72] and indeed, the fact that models make different assumptions about cloud formation explains much of the variations in climate sensitivity predictions between models.[73] Despite the uncertainties, climate modelers must make decisions about how to represent these processes in their models, how to interpret data or assign probabilities to hypotheses about future climate change, and how to integrate data from physical science models to investigate the impacts of climate change. Such decisions depend on determining what risks of error are acceptable, which in turn depends on ethical judgments about how *bad* the consequences of error will be.[74]

But, insofar as scientific decision making involves social and ethical values, then defining expertise in narrow terms invests scientists with sole authority to make value judgments that affect a variety of stakeholders. It also illegitimately excludes members of the public from having a say in scientific areas that have significant effects on people's well-being.[75] Indeed, not only do many nonscientist stakeholders possess knowledge and interests relevant to evaluating and endorsing particular value judgments involved in scientific research but also there are no good reasons to believe that scientists *qua* scientists either have such expertise or are sufficiently diverse to identify or evaluate the plurality of relevant interests. Indeed, for this reason, there have been increasing efforts to incorporate stakeholder input throughout the research process in areas where values play a role, such as in climate research.[76]

As with shared standards, then, we can interpret expertise in more or less robust ways. Understood "thickly," being an expert might require formal training and a significant number of peer-reviewed publications on the exact topic related to the research being evaluated. In this case, only a small number of scientists from a specific discipline narrowly construed will count as qualified to advance normatively appropriate dissent. On this account, we would be able to capture at least some of the cases that many take to be paradigms of NID, such as climate change. As indicated earlier, many of the most prominent climate skeptics are not, for example, climate modelers nor do they have peer-reviewed articles published on climate modeling.[77] Similarly, Jenny McCarthy's proud proclamation that she received her degree from "The University of Google" would also clearly render her dissenting views about the purported link between the MMR vaccine and autism as normatively inappropriate.[78] However, as we have seen, it is not clear that ID dissent could be classified as normatively inappropriate under this criterion. More problematically, a thick interpretation of expertise would render much epistemically valuable dissent as normatively inappropriate. Alternatively, the requirement for expertise might be interpreted more thinly, as perhaps involving some relevant experience that bears on the topic under consideration. This interpretation would allow dissent originating from scientists from various disciplines and from informed, nonformally trained scientists, but such understanding would not clearly rule out those cases that many take to be problematic instances of dissent.

Of course, none of this is to say that scientific expertise is irrelevant either to evaluate scientific claims or to make policy or behavioral decisions. It is reasonable to take scientific expertise as pertinent when determining how much weight certain claims should receive in evaluating the evidence for or against a particular hypothesis. Scientific experts make important contributions to society, and thus their intellectual authority should be appropriately acknowledged. If one assumes that a truth-tracking process has been followed, for instance, then the fact that virtually all climate scientists agree that anthropogenic global warming is occurring is highly compelling evidence to accept this claim as true, or justified. Our point is merely that there are dangers in defining expertise narrowly for purposes of discouraging, suppressing, ignoring, or refusing altogether to engage with some dissenting views.

Conclusion

While it seems reasonable to think that requirements to "play by the rules" constitute a good criterion to identify NID, we have shown in this chapter that it is not easy to make judgments about what exactly this means. The criterion of uptake is useful only relative to a set of shared standards. The shared standards criterion can be interpreted more thickly as requiring significant agreement among participants about the number of standards that need to be shared and the understanding of such standards, or more thinly as demanding minimal accord. Along this spectrum, the thicker the condition, the greater the risks of excluding epistemically valuable dissent, and thus of limiting the valuable role that dissent plays in knowledge production. The thinner the condition, the greater the risk of identifying as appropriate dissent that actually fails to advance or can hinder scientific progress. Similarly, expertise can be interpreted in more or less strict ways. Either way, the criterion fails to reliably identify NID by either capturing cases of dissent that can actually promote knowledge or excluding cases that fail to do so.

Notice that the arguments presented here do not show that the criteria discussed are useless in identifying as appropriate or inappropriate particular instances of dissent. It might well be the case that some cases of dissent clearly and uncontroversially fail to adhere to any or all of the criteria discussed, and thus are easy to identify as normatively inappropriate. Our argument, however, is that they cannot do so *reliably*. Depending on how the criteria in question are understood, they will result in either classifying as appropriate some dissent that fails to play the roles for which dissent is valuable or identifying as inappropriate some dissent that can in fact contribute to knowledge production.

Importantly also, even if these criteria do not provide us with a reliable way to identify NID, they can still be valuable for many other purposes. Longino, for example, has argued that sharing standards, uptake of criticism, and expertise—or tempered equality of intellectual authority—constitute criteria for the objectivity of scientific communities.[79] Such criteria, for example, may reliably indicate the extent to which a scientific community is appropriately organized and functioning so as to minimize bias. Sharing standards may also be important for understanding when we can say that a scientific theory is justified over alternatives. Ability to engage in uptake, while difficult to judge precisely, may be an important virtue to cultivate within individual scientists. Similarly, particular conceptions of expertise, more or less robust, may be important for making some determinations, such as what literature should be included in a policy advising report or who should be asked to review a particular article or serve on a scientific advisory committee. Yet while the notions of shared standards, uptake, and expertise may be useful concepts for a variety of purposes, we conclude that they cannot serve as criteria that reliably identify NID.

5

Imposing Unfair Risks

In January 2017, NASA and the National Oceanic and Atmospheric Administration announced that 2016 was the hottest year on record since scientists began tracking the average global temperature 100 years ago (NASA 2017). Indeed, it is the third year in a row that the heat record has been broken.[1] An increase in wildfires attributed to anthropogenic climate change have burned an additional 16,000 square miles of land in the United States—equivalent to the size of Massachusetts and Connecticut combined.[2] Scientists now predict that Arctic Sea ice will be nonexistent during the summer months as early as 2020,[3] which will contribute to the rising of sea levels by six feet by the end of the century, cause the loss of several Pacific Islands, and threaten 13.1 million people in the United States with coastal flooding.[4]

Climate change dissent challenges these findings. But in doing so not only does it contest certain empirical claims but it also imposes risks. Climate change skepticism, like all dissent involving policy-relevant science, is risky. Erroneously accepting dissenting claims that turn out to be false can result in harm to health and quality of life, loss of life, destruction of property, and extinction of cultures. Moreover, in many instances, such risks fall largely on the public in general and on those who are the least well off in particular. At the same time, erroneous acceptance of climate change dissent tends to benefit powerful and wealthy companies that may suffer economic losses from potential regulatory policies. In addition, dissent can result in an unlevel playing field, where some dissenters who stand to benefit from rejecting a consensus view have significantly more power and resources to conduct research and disseminate their views. In this context it is important to consider whether dissent that imposes unfair risks on the public can hinder scientific progress and thus constitute normatively inappropriate dissent (NID).

In a recent proposal, Justin Biddle and Anna Leuschner draw upon the notion of inductive risk[5]—that is, the risk of wrongly accepting or rejecting a hypothesis—to offer a set of jointly sufficient conditions they believe can be used to identify NID dissent.[6] The conditions are: (1) the dissent's real-world consequences are severe; (2) it violates established conventional standards; (3) it protects industries and producers while putting the public at risk; and (4) producer and public risks fall largely upon different parties. They contend that current organizational and regulatory aspects regarding the production and dissemination of scientific research make it likely that dissent meeting these conditions will impede knowledge production and therefore constitute epistemically detrimental dissent.

To be clear, theirs is not a strategy attempting to identify *necessary and sufficient* conditions for NID. Rather, as indicated earlier, their goal is to ascertain *jointly sufficient* conditions that can be used to identify dissent likely to hinder knowledge production.[7] Yet, if successful, this would still enable us to recognize some epistemically harmful dissent. This would be helpful in limiting the damage resulting from instances of NID, even if the account excludes cases of dissent that result in other adverse consequences, such as public confusion or stalled policies, or that fail to promote scientific progress but that do not actually impede it. Furthermore, it can be a useful account because it considers the actual context in which science is being produced, rather than providing criteria that only apply in idealized circumstances. Indeed, at least part of the motivation behind Biddle and Leuschner's proposal is the recognition— shared by many—that the context in which modern science is produced is very far from ideal.

In this chapter we evaluate this inductive risk account of NID and, in spite of its promise, we find it wanting. We begin by discussing the notion of inductive risk. We then present and critically evaluate the inductive risk account (IndRA) developed by Biddle and Leuschner. We argue that the proposed characteristics fail to reliably identify NID. This is the case in part because the IndRA relies on the criterion of shared standards, a criterion that, as we saw in chapter 4, is not straightforward. Moreover, because of the ambiguities present in judgments about inductive risks, this account turns out to be of little use in practice.

The Inductive Risk Account of NID

Accepting and rejecting scientific hypotheses invariably carries some degree of uncertainty and thus there is always a chance that science will get it wrong. The risk of falsely accepting a hypothesis or theory is known as "inductive risk." In deciding to accept or reject a scientific hypothesis, then, scientists must consider whether there is enough evidence to do so. How much evidence is needed depends in part on how serious the potential consequences of error are, or on decisions about what types of risks are more acceptable. When policymakers use science as the basis for public policy, accepting or rejecting a consensus view can carry rather serious risks of error. For example, if scientists mistakenly accept a hypothesis that a certain level of a chemical substance is harmful to human health, this can lead to overregulating the chemical in ways that impose social and economic costs on the producers of the substance in question, on their shareholders, and ultimately on the consumers of products containing such substance. On the other hand, if scientists erroneously reject such a hypothesis, the substance in question is likely to be underregulated, which will impact the health of many individuals, as well as result in associated social and economic costs to the health-care system. Determining when there is sufficient evidence to accept or reject a hypothesis, then, requires deciding which risks are more acceptable. This involves weighing the various ethical and social interests at stake.

Given the need to make inductive risk value judgments in science and policy decisions, Biddle and Leuschner provide a set of jointly sufficient conditions to identify dissent that imposes unfair risks and, thus, is NID.[8] According to the IndRA, research, and its dissemination, dissenting from a consensus hypothesis H is normatively inappropriate when each of the following conditions obtain:

1. The nonepistemic consequences of wrongly rejecting the consensus hypothesis H are likely to be severe.
2. It violates shared standards.
3. It involves intolerance for producer risks at the expense of public risks.
4. Producer risks and public risks largely fall on different parties.

Biddle and Leuschner focus on challenges to the consensus view in climate change research in order to illustrate how their criteria account for what is problematic about such dissent. As we have seen, evidence shows that the overwhelming majority of climate scientists agree that humans are causing global warming.[9] But as already mentioned, this consensus has been insistently challenged by some researchers.[10] Does this dissent meet the IndRA conditions?

Biddle and Leuschner believe it does. Indeed, most would grant that condition no. 1 is easily satisfied in the case of climate change science. As with most policy-relevant science, the nonepistemic consequences of falsely rejecting the existence of anthropogenic global warming can be severe, from inaction on the side of the public and policymakers to an increase of the behaviors— for example, use of fossil fuels— that are responsible for the raising in global temperatures. Without the enactment of policies to provide substantial mitigation of current fossil fuel practices, researchers estimate that there will be a 3.7 to 4.8°C increase in global mean surface temperatures by 2100 compared to pre-industrial times.[11] Adverse impacts that can result from global warming go from melting ice and rising sea levels resulting in flooding and erosion of coastal and low-lying areas, to more frequent droughts and severe wildfires, leading to loss of agricultural land and consequent regional food insecurity, to increased health problems and severe economic costs to individuals and society. Should temperatures increase beyond 4°C, scientists predict that the results will be even more devastating, with substantial species extinction, more widespread food insecurity, and limited potential for adaptation in some areas.[12]

According to Biddle and Leuschner, at least some of the research that dissenters have used to challenge the consensus on climate change fails to adhere to shared standards in science, and thus climate change dissent also meets condition no. 2 of the IndRA. For instance, they consider challenges to the "hockey-stick graph" presented by climate scientists Michael Mann, Raymond Bradley, and Malcolm Hughes,[13] which provided a compelling visual of the significant impact of anthropogenic CO_2 equivalent gases on average global temperature. The hockey-stick graph reconstructed the average global temperature between 1400 and 1980 using data from tree rings, coral deposits, ice cores, and other data to act as proxies for temperature. The graph showed unprecedented warming that began to occur with the human industrial revolution (when there was a significant increase in greenhouse gases) and continues to rise dramatically afterwards. Stephen McIntyre and Ross McKitrick criticized this graph by claiming that it depended on flawed tree ring data, which had the effect of underestimating the temperatures in the past, particularly in the fifteenth century.[14] Indeed, by eliminating large sets of tree data on which Mann, Bradley, and Hughes relied in their study, McIntyre and McKitrick produced a different graph showing *higher* sustained average temperatures in the 1500s than in the twentieth century.[15] Climate change skeptics took this as support for the view that twentieth-century average global temperature increases were the result not of human activity, as Mann, Bradley, and Hughes's work purported to show, but of natural fluctuations in the climate system.

However, many have found McIntyre and McKitrick's exclusion of certain proxy data, as well as the preprocessing algorithms they used to distinguish climate signal-to-noise in the proxy data, unjustified.[16] By using methods that classified relevant data as "noise," they excluded data pertinent to reconstructing temperatures in the 1500s, which made current global temperature increase appear as part of a natural fluctuation pattern.[17] Biddle and Leuschner argue that by failing to ensure that the data they had ignored was actually irrelevant,[18] McIntyre and McKitrick's research violated conventional standards used in climate research. Indeed, the central findings of Bradley, Mann, and Hughes in their reconstruction of the temperature record from 1400 to 1980 have now been independently reproduced and corroborated by a variety of studies.[19]

Although Biddle and Leuschner recognize that the case of climate change research is complex, they also believe that this dissent involves an intolerance of producer risks at the expense of public risks, and that these risks fall generally on different parties. On the one hand, the consequences of wrongly accepting the consensus hypothesis fall primarily on the fossil fuel industries and their shareholders, as such industries would be unnecessarily regulated and the use of their products needlessly restricted with the resultant loss of profits. On the other hand, the consequences of wrongly rejecting the consensus hypothesis fall primarily on the public, who will suffer the health, environmental, and economic effects of sea-level rise, floods, and droughts. Thus, Biddle and Leuschner argue that climate change dissent also meets conditions nos. 3 and 4 of the IndRA.

According to their analysis, then, the IndRA shows that research dissenting from the anthropogenic global warming hypothesis, as well as its dissemination, constitutes NID. Now, why should one think that an unjust distribution of risks constitutes a criterion for NID? Indeed, why should the distribution of risks matter at all to the epistemic value of dissent? Most would agree that research that leads to an unfair risk distribution constitutes an ethical problem, but why would such characteristic be of relevance to whether the dissent in question contributes or hinders knowledge production? Biddle and Leuschner claim this is the case because of the current context in which science is conducted. The IndRA recognizes knowledge production as occurring within particular organizational and regulatory practices, and it is within these current practices that they believe some dissent—dissent meeting the IndRA conditions—hinders knowledge production. In a different context, such dissent might well not have such an effect on scientific progress.[20]

What is it, then, about our current social and political context, particularly in countries like the United States, that make the IndRA a plausible one to identify NID? The conduct and, particularly, the dissemination of research are highly influenced by political, economic, and ideological interests, Biddle and Leuschner argue. Industry's direct and indirect influence over science has been increasing, both through its ability to fund research and its power to disseminate results. Indeed, as is well known, industry accounts for the majority of funding in scientific research.[21] For instance, in 2014, the United States invested approximately $465 billion, about 2.8% of its gross domestic product (GDP), in research and development (R&D).[22] Two-thirds of this R&D spending came from industry-funded R&D, about one-fourth was federally funded, and the remainder came from academia, nonprofits, and other funders.[23] Pharmaceutical and medical-device companies now supply 58% of total biomedical research spending,[24] and contribute more than six times the amount of funding for U.S. drug trials than the federal government.[25] The tobacco industry has a long and well-documented history of funding scientific research.[26] It has employed prominent academics as researchers, as well as provided extensive funds to universities. Much of its funded research has focused on confounding variables such as stress, diet, and genetic predispositions as a way to undermine the link between smoking and chronic disease. The food industry is also increasingly funding more research. For instance, a recent disclosure by Coca-Cola indicated that it has spent nearly $119 million in the past five years on scientific research and on health and well-being partnerships.[27] The disclosure also shows that Coca-Cola has given millions of dollars to several influential medical organizations, including the American Cancer Society, the American College of Cardiology, and the Academy of Nutrition and Dietetics.[28] The fossil fuel industry also invests great amounts of money in research.[29]

Given the substantial investment that a variety of industries make in scientific research and its dissemination, it is unsurprising that they exert significant influence over various aspects of scientific research.[30] Financial interests can have crucial effects on what research problems to address and how to frame such problems.[31] But they can also influence the selection of evidence, interpretation of data, methodological choices, and the reporting of results.[32] Moreover, industry exerts significant influence over whether and when research is made public.[33] Consider, for instance, recent reports that the fossil fuel industry had knowledge regarding the impact of their products on climate. According to newly released evidence, Exxon scientists conducted cutting-edge climate research during the late 1950s and warned top management of the potentially catastrophic risks posed by global warming. The oil industry, however, not only failed to chart a different course in light of the results of its own research but also failed to make the information public.[34]

Of course, in a context where financial incentives are powerful, economic interests can contribute to creating an environment in which research likely to have a negative impact on such interests can be restricted while projects likely to support them can flourish.[35] On the one hand, researchers conducting investigations favorable to certain industries are likely to be rewarded with funding for their work. On the other hand, those who conduct research that can conflict with industry's interest can be silenced, their reputations tarnished, and their funding put in jeopardy.[36] These were well-known strategies of the tobacco industry, which both funded research that attempted to prove that tobacco products posed no significant health effects and suppressed dissenting voices by mounting serious attacks on the careers of scientists.[37] Biddle and Leuschner's contention is that their IndRA captures characteristics that, in the current social and political context of research production, can identify NID.[38] In this context, conducting and disseminating dissenting research that skews the distribution of risks toward those who have more power can result in attacks to scientists who defend a consensus view inconsistent with industry's interests, such as the consensus on global warming. These confrontations can be severe and burdensome. Dissenting voices can create an environment of intimidation so that it steers scientists away from particular fields of inquiry, drives them to change their projects, or leads them to defend their hypotheses less forcefully than they otherwise believe is appropriate.

Such dissent can also lead to epistemic corruption—that is, to the creation of social conditions conducive to the development of epistemic vices by agents operating within them.[39] These actions can thus ultimately slow down or impede scientific progress in particular scientific areas. Moreover, in this context, dissenting voices can be used to force scientists who support the consensus to spend time and resources defending both themselves and their investigations, which in turns prevents them from dedicating time to their own research, again slowing down the production of knowledge.

Yet, it is worth noting that the IndRA fails to capture instances of dissent that others have thought to be normatively inappropriate. For instance, as we have seen, many take both dissent about the relationships between AIDS and HIV[40] and dissent regarding the safety of childhood vaccines[41] as paradigmatic instances of NID.[42] Nonetheless, the IndRA excludes both of these instances of dissent because they are not cases involving an intolerance for producers' risks at the expense of the public's risks. Indeed, in the case of pediatric vaccines, dissenters are skeptical about their safety at least in part because they fear that vaccine producers are hiding risks to public health in order to maximize profits.[43]

Of course, the exclusion of these cases need not present a decisive challenge to the IndRA. As mentioned, the account only intends to provide jointly sufficient conditions for NID, but not necessary ones. Hence, although these other cases of dissent fail to meet the IndRA conditions, this does not mean they do not constitute cases of NID. Furthermore, as indicated earlier, the IndRA is concerned with dissent that slows down or hinders scientific progress. It might be, then, that although these other cases of dissent can result in adverse nonepistemic consequences—for example, negative impacts to public health—they do not affect scientific progress. Be that as it may, insofar as the goal is to provide criteria that can reliably identify dissent likely to result also in severe nonepistemic consequences whether or not it hinders scientific progress, that the IndRA cannot capture these cases might be a problem. After all, if the goal is to reduce or eliminate adverse consequences of dissenting views, then the inability to identify as normatively inappropriate cases of dissent that are detrimental in ways other than epistemic ones would constrain our ability to achieve such a goal. In what follows, however, we put this concern aside.

Evaluating the Inductive Risk Account

Its insights notwithstanding, the IndRA is also unsuccessful in reliably identifying NID. This is the case even granting its dependence on a particular social context. Indeed, as indicated earlier, its focus on contextual factors can be thought of as a strength of this account insofar as it considers nonideal conditions of knowledge production.[44] Nonetheless it suffers from various flaws.

Let's begin with condition no. 1—that is, the requirement that the nonepistemic consequences of wrongly rejecting the consensus hypothesis H are likely to be severe. We take this to be a threshold condition for determining whether a particular instance of dissent is normatively inappropriate and thus will likely result in negative epistemic consequences. That is, if nonepistemic adverse effects do not follow from a dissenting view, the other conditions need not be of concern, as such dissent will presumably fail to slow or hinder knowledge production. As Biddle and Leuschner note,[45] dissent from the Einsteinian hypothesis that simultaneity is relative is unlikely to have any significant negative social consequences and thus is not of concern. They argue that even assuming that dissent from such hypothesis involves the violations of shared standards, the dissent will not be epistemically detrimental, and thus does not involve a case of NID, because scientists defending the consensus will not be particularly affected by the dissenting views; they can continue their research as they see fit, and they will not be afraid of defending their consensus views as forcefully as they believe is suitable.

This seems correct. When the dissent is related to scientific areas that are unlikely to affect public behavior, or impact public policy, and where political and economic interests are inexistent or minimal, it is improbable that such dissent will hinder scientific progress. Nonetheless, determining what the consequences of falsely rejecting a consensus view might be or how to assess their severity is not always an easy task.[46] Consider, for instance, the hypothesis H: "Increasing average global temperature will cause extreme weather events with greater frequency." What would the impacts of falsely rejecting H be? A shift in adaptation priorities could result and this might have negative consequences for many stakeholders, including the destruction of property, lives lost, an increase in certain diseases, and economic and environmental destruction of agricultural areas. Nonetheless, the adverse effects on various stakeholders are bound to be different and thus the assessment of their severity is difficult. For example, those living in resource-poor areas with deficient infrastructure are more likely to experience worse consequences than those who already have adequate infrastructure that offers more protection against high winds or flooding. Furthermore, many communities, concerned about the potential impact of extreme weather events, might continue to demand resources to prepare for flooding in low-lying areas (regardless of whether climate change would be the cause), thus limiting the negative consequences of falsely rejecting H.

Of course, in many cases related to policy-relevant science, determinations that rejecting a consensus view is likely to have severe nonepistemic consequences are reasonable, and when this happens, then attention to the remaining conditions of the IndRA becomes relevant in order to determine whether the dissent in question is normatively inappropriate. Unfortunately, such conditions suffer from a variety of problems. Take the need for shared standards. First, it is not completely clear what Biddle and Leuschner have in mind regarding what constitutes "shared standards." Some of Biddle and Leuschner's discussion seems to indicate that they are referring to simple methodological rules—for example, that scientists should choose animal models based on responsiveness to endocrine-active agents of concern, not on convenience and familiarity;[47] or that they should ensure the data they toss out is noise and not signal.[48] In some of their work they also suggest that false or misleading characterizations of what opponents have done or said also constitutes a violation of the shared standards condition.[49]

Clearly, insofar as both proponents of the consensus and supporters of the dissent agree that a particular shared rule has not being followed, then this condition can be used to identify some dissent as normatively inappropriate. But not uncommonly, dissenters and consensus supporters can disagree, not about the methodological rule but about whether the rule in question has been used appropriately—for instance, what one group identifies as noise, the other interprets as signal. For example, McIntyre and McKitrick have argued that their exclusion of certain bristlecone pine tree ring data was warranted.[50] It is true that an assortment of variables influences whether tree ring data at certain sites is more or less reliable as an indicator of temperature. Although temperature is typically a significant cause of wood pulp growth, this can be affected by other factors such altitude, increases in CO_2, or changes in light or precipitation.[51] Indeed, in 1999, when the IPCC Working Group was deciding whether to give a prominent spot to Mann and colleagues' hockey-stick graph in the IPCC *Third Assessment Report: Summary for Policymakers*, there was significant debate among climate scientists about whether the inclusion or exclusion of certain proxy data was justified. In emails, Keith Briffa, for example, disagreed with Mann's proxy data judgments and urged caution about featuring the hockey-stick graph in the IPCC report:

> It should not be taken as read that Mike's series is THE CORRECT ONE. I know there is pressure to present a nice tidy story as regards 'apparent unprecedented warming in a thousand years or more in the proxy data', but in reality the situation is not quite so simple.[52]

Briffa later largely came to agree with Mann's findings, but the methodological disagreements at the time appear to have been reasonable.

To be sure, in some cases it might be clear that a claim about whether particular data points are noise or about whether a standard methodological convention applies is wrong, but, again not uncommonly, such determinations will depend on background assumptions that not all actors will share, even if they share the methodological rule in question. What some scientists will see as failing to share a standard, others will see as being cautious and requiring "better" evidence before racing into costly environmental or health policies. Hence, such disagreements may reflect reasonable disputes about what epistemic tradeoffs will best advance both the epistemic and the policy aims of the research. For example, clinical researchers investigating a new drug might agree that randomized controlled clinical trials are the most scientifically rigorous method of hypothesis testing. Nonetheless, they can disagree about the selection of a comparison group. Such disagreements might stem from different interpretations regarding the evidence for what constitutes the standard of care at the moment of trial design. For instance, when researchers began to design trials to add new drugs to existing regimens against ovarian cancer, disagreements arose among the collaborating international teams of trialists about the appropriate control arm. While some of them considered that prior evidence had shown the addition of taxanes to be effective, and argued that the control arm should consist solely of the taxane-containing regimens, others thought such evidence was lacking and argued for flexibility in the control group.[53]

Additionally, as we saw in chapter 4, scientists might agree on particular methodological rules but disagree about other relevant standards, and presumably the sharing of such standards might also affect the outcomes of particular research inquiries. For instance, dissenters and supporters of the consensus can differ on their commitments to particular epistemic values, such as empirical adequacy, consistency with other accepted theories, expansion of knowledge, or explanatory power.[54] They may also disagree about social values important to the evaluation of a hypothesis, such as how well the hypothesis in question is able contribute to the satisfaction of particular social needs.[55] Of course, in some cases, dissenters might claim to share standards and to apply them diligently when they are, in fact, flouting the standards in question.

Nonetheless, as discussed in the prior chapter, determining what counts as a shared standard is not clear-cut and requires a tradeoff between preserving the epistemic benefits of dissent and limiting the negative effects of NID. On the one hand, too thick an understanding of shared standards risks categorizing some epistemically beneficial dissent as normatively inappropriate, and thus undermining epistemic pluralism and the important beneficial role that dissent can play in enhancing objectivity and increasing our confidence in a consensus view. On the other hand, a thinner conception of shared standards can preserve epistemic benefits of pluralism at the cost of failing to identify some NID that will turn out to be epistemically detrimental. Determining whether or when the condition of shared standards is met is thus not an easy task.

Perhaps, the conditions regarding the distribution of risk—that is, conditions nos. 3 and 4 of the IndRA—can actually be used as a way to assess the appropriate tradeoff between epistemic pluralism and problems resulting from NID. That is, one could argue that the epistemic benefits of particular instances of dissent are outweighed when conditions nos. 3 and 4 are met. Under this interpretation, determinations about whether researchers have violated shared standards or not—or perhaps about whether the violations are relevant to the normative status of the dissent in question—will depend on judgments about the fair distribution of risks related to falsely rejecting a consensus view. When there is an unfair risk distribution, we have more reasons to require a thicker conception of shared standards.

Unfortunately, although this might appear initially plausible and helpful, these conditions are also difficult to assess, and thus are unlikely to offer much help in identifying normatively inappropriate dissent. First, conditions 3 and 4 attempt to identify dissent that is not just *risky*—that is, dissent that has severe nonepistemic consequences—but also that unjustifiably protects the interests of producers at the expense of the public's interests. It is true that in some cases it might be relatively easy to determine whether these condition are met. Take, for example, dissent regarding tobacco safety. On the one hand, it is reasonable to argue that falsely rejecting the consensus hypothesis that smoking is dangerous to human health would have serious adverse consequences for public health. Falsely accepting the consensus view, on the other hand, would primarily negatively impact producers, as their product would be unjustifiably restricted and they would suffer loss of profit unnecessarily Therefore, one could uncontroversially argue that dissent regarding the safety of tobacco products clearly involves intolerance for producer risks at the expense of risks to the public, where these risks fall largely on different parties.

Nonetheless, although clear-cut inductive risk judgments might be easily identified in some cases, this is unlikely to be so in most policy-related science, where such judgments are often complex and involve a variety of risks to various parties. Climate change dissent, which Biddle and Leuschner take to be a paradigmatic case of epistemically detrimental dissent, makes these difficulties clear. Consider the consensus hypothesis H: "Average global temperature will increase by 2°C over the next 100 years." Arguably, there are risks to both producers and the public in falsely rejecting H. It is true that falsely accepting H may lead to overregulation and loss of profits for oil companies, but it could benefit for-profit producers of alternative energy sources, such as wind or solar power. It could also benefit nuclear power companies. Thus, any particular instance of dissent in complex cases is unlikely to involve an intolerance of risk to *all* producers. Should the interest of some producers count as more important than those of others?

Similarly, the "public" is not a monolithic group. Especially in a case such as climate change, many different stakeholders with diverse and conflicting interests constitute the public or publics. Hence, falsely rejecting H will certainly have negative consequences for some members of the public, particularly those in resource-poor countries that are most likely to be affected by climate change. However, it may benefit some stakeholders in the Global North who are likely to experience fewer impacts from climate change and will suffer the negative economic effects of regulating CO_2 emissions more strictly. Similarly, while future generations might be significantly worse off if H is falsely rejected, people now alive will be less affected by such rejection and more so by its acceptance. Therefore, in complex cases, such as climate change dissent, it would be challenging to determine whether conditions 3 and 4 are met, as the dissent could involve intolerance for some producer risk only but not for others and tolerance for risk to some members of the public but not to others.[56]

If this assessment is correct, climate change dissent does not clearly constitute a case of NID on the IndRA. Of course, that climate change dissent does not obviously meet the conditions identified by the inductive risk account is not sufficient to reject such an account. After all, it could be that Biddle and Leuschner have misidentified climate change dissent as normatively inappropriate when is not so (as counterintuitive as this might seem). Nonetheless, given that many consider climate change dissent to constitute a clear case of inappropriate dissent, the fact that the IndRA does not capture this dissent appears problematic. Moreover, the authors use climate change dissent as a paradigmatic instance of epistemically detrimental dissent.

Conclusion

Although the IndRA might appear a promising strategy for reliably identifying normatively inappropriate dissent, we have shown here that the promise is likely to go unfulfilled. This is not because we believe that ultimately even bad dissent can have epistemic benefits. Indeed, we grant that some dissent could indeed do more harm than good. The reason why the IndRA account is unsuccessful is that it fails to reliably identify when that is the case. This is so because, as we have argued, it is sometimes difficult to assess whether falsely rejecting a consensus hypothesis will have adverse consequences, as well as to determine how severe such consequences will be. Even when such assessments are possible, evaluations about whether the risks of such false rejection would fall primarily on the public while benefiting producers are far from clear-cut in many cases of policy-relevant science. Using these criteria to identify normatively inappropriate dissent is unlikely to be particularly useful in practice or be particularly reliable.

Moreover, the IndRA involves only jointly sufficient conditions for normatively inappropriate dissent. Although this is not a decisive argument against the view, it is a problem insofar as a main goal of providing criteria to identify NID is not only to prevent adverse epistemic damages that might result from such dissent but also to identify dissent that offers no epistemic value and can produce negative social effects. After all, there can be cases of NID that do not involve an unjust distribution of risks, but that nonetheless have severe epistemic and social adverse consequences.

This is not to say that the IndRA does not provide important resources for philosophers of science. For example, the IndRA may be useful in helping us explain, after the fact, the reasons why some dissent has led to epistemically detrimental consequences. One could argue that climate change skepticism is harmful in the United States because there is an uneven playing field where dissenters have disproportionate political power and they tend to prioritize minimizing producer risks. In such contexts, scientists are likely to feel intimidated and err on the side of caution when they ought not to do so. Yet in order to provide such an explanation, one would have to beg the question that is of central concern to us here. To claim that some dissent has the bad consequence of hindering knowledge production—that is, that some dissent is epistemically detrimental—is to presuppose that the dissent in question is in fact normatively inappropriate, or bad. Thus, we need independent criteria for reliably identifying some cases of dissent as normatively inappropriate. Clearly, without assuming that climate change dissent is normatively inappropriate, it is difficult to see how one could claim that it results in epistemically negative effects. For if such dissent were in fact true, the fact that it leads scientists to be conservative rather than overly bold or alarmist in their pronouncements would not be epistemically detrimental. As we mentioned in chapter 1, judging that the effects of certain instances of dissent are in fact negative presupposes that the dissent in question is normatively inappropriate. Thus, such an account is first needed.

Our review of the various criteria that have been proposed to reliably identify NID shows that finding such criteria proves to be a difficult task. Without denying that some dissent might indeed be normatively inappropriate, it might then be best to jettison such attempts and focus on the conditions that make it likely some dissent will have adverse epistemic and social consequences. We turn to that task in the following chapters.

6

Dealing with Normatively Inappropriate Dissent

Evidence shows that a significant amount of the public harbors misunderstandings regarding the extent of the scientific agreement on climate change,[1] as well as the evolution of human beings.[2] Studies have also reported false beliefs about the existence of agreement in the scientific community regarding the safety of the MMR vaccine as a result of dissenting views highly publicized by the media.[3] Normatively inappropriate dissent (NID) can thus lead to confusion and false beliefs about the existence of a scientific consensus or the reliability of the evidence.

Moreover, given the epistemic obligations that scientists have to engage dissenting views,[4] NID can slow or impede scientific progress by forcing scientists to waste time and resources responding to it.[5] In a context where dissenters have significant economic and political resources, scientists can feel intimidated, and can refrain from working on or publicizing research that threatens powerful financial and ideological interests.[6] Similarly, scientists who support a consensus view might become more tentative or conservative in their claims than warranted because of fears of negative repercussions, which can produce biased results.[7]

In cases where the scientific evidence is relevant for policy or action, these negative epistemic consequences can also have serious adverse social impacts. To the extent that NID appears to challenge the scientific consensus, it can lead the public and policymakers to question the strength of the evidence supporting adoption of particular public policies or behaviors. Lack of support, or outright opposition, among many members of the public regarding measures to address climate change impacts or vaccination policies exemplify the social damages of NID.[8] Indeed, the refusal of the Mbeki's administration to allow the use of antiretroviral drugs against HIV, with the consequent deaths and HIV infections, makes the costs of NID harrowing.

The use of both time and funds required to respond to NID can also result in social harms because of opportunity costs. The time and money that scientists need to employ responding to epistemically useless dissent cannot be utilized for other, more productive research purposes that would be of benefit to the public. Similarly, conservatism in presenting scientific results is problematic not only for its epistemic damages but also for its social ones. If, for example, the full extent of a natural phenomenon such as global warming is unacknowledged, public policies intended to prevent or limit the damages are likely to be inadequate.

As we have seen in chapters 3 to 5, finding criteria that can reliably identify NID is not an easy task. Bad-faith motives, failure to play by the rules, and the unfair imposition of risks all fail in this regard. We believe that such criteria would need to be reliable if one wants to use them to identify dissent that should be discouraged, suppressed, or ignored in some way. This is so, because the epistemic and social benefits of dissent are significant and the risks of losing these benefits are serious. Of course, that the criteria discussed in prior chapters are ineffective in reliably detecting NID does not mean that scholars cannot find successful criteria. But would the ability to reliably identify NID help to limit or prevent its negative effects? In what follows, we grant, for the sake of argument, that scholars can find some alternative criteria that would reliably track NID. We then consider a variety of ways in which stakeholders could use such identification so as to minimize the epistemic and social damages that such dissent can inflict, including prohibiting the dissent in question, targeting it for special scrutiny, placing limits on scientists' epistemic obligations, guiding public beliefs, emphasizing the existence of a consensus, and discrediting dissenters. We evaluate the extent to which these strategies could be effective. We show that some of the strategies, such as prohibiting NID, could prevent the damages produced by NID, but its implementation—at least in democratic societies—is unlikely, for practical and ethical reasons. Others, such as imposing limits on scientists' epistemic obligations to engage dissenting views, are unhelpful in addressing the problems mentioned and may even exacerbate them. Other strategies, such as emphasizing the consensus, could be effective, but are likely to generate equally serious problems. Moreover, the existence of criteria that can reliably identify NID is not sufficient to ensure that the dissent in question, and only NID, is classified as such. Those using such criteria need to also apply them correctly. How effective such criteria can be in addressing problems resulting from NID will thus also depend on how well different stakeholders, from laypeople to policymakers, can use those criteria in practice. This will obviously depend on how much expertise is needed to correctly employ the relevant criteria. For these reasons, we argue that finding criteria to reliably identify NID would be of limited use in helping address the adverse epistemic and social consequences such dissent creates.

Prohibiting NID

An obvious strategy that one could implement as a response to NID would be to prohibit the dissent in question. The prohibition could focus on the dissemination of dissenting views. Sources of diffusion, such as academic journals and publishers and the media, could simply ensure that NID is not published or that it does not receive media attention. This strategy might go a long way in trying to address some of the most troubling concerns about NID. Prohibiting NID could prevent or assuage public confusion, lack of support for needed policies, wasted resources, and unwarranted conservatism in scientific results. Nonetheless, in democratic societies this strategy is implausible. It clearly conflicts with democratic principles of free speech and expression. Moreover, there are no mechanisms in place that could conceivably prevent dissenters from disseminating their views via the Internet and other open publishing sources.

Prohibition strategies could alternatively focus on severely restricting, or outright eliminating, funding for research that supports NID.[9] For example, public granting agencies could give low funding priority to research conducted in support of NID. Insofar as the research is not conducted, this would limit dissemination. Nonetheless, although this strategy might appear more feasible and less fraught with ethical concerns, it is unlikely to be particularly effective in preventing the negative effects of NID. Dissenting research is largely privately funded, and therefore putting limits on the use of public resources would make only a small dent in its production. For instance, evidence indicates that between 2003 and 2013, ninety-one conservative/free-market organizations spent a combined annual budget of $900 million in order to produce and disseminate dissenting views on climate change.[10] Similarly, the alcohol, tobacco, and food industries have spent, and continue to spend, millions of dollars funding research that understates the harms of their products and creating professional societies, research centers, and charities to disseminate the conclusions of such research.[11]

A further strategy to limit the dissemination of certain dissent without flat-out banning it would be to have search engines such as Google "flag" highly questionable, unsubstantiated, or discredited ideas. Wikipedia, for example, allows users to "tag" entries that may have incorrect content. This can alert both editors and readers that the entry may not be reliable. Yet there are often disputes over the tags themselves that can be difficult to resolve without getting into a debate about the content of the entry. Moreover, as editors of Wikipedia have noted, this strategy can also lead to "tag-bombing" where entries are indiscriminately tagged or challenged simply to be disruptive or make it difficult for users to evaluate whether particular tags or warnings are warranted.

Targeting NID for Criticism

The ability to identify NID could also be helpful in determining that scientists and others should more rigorously scrutinize and challenge the dissent.[12] The scientific community could make a concerted effort to expose shady evidence, reveal biases, and pinpoint suspect methodological practices that could limit public and policymakers' misperceptions and correct false beliefs about the strength of the evidence. Limiting misperceptions could also address concerns about failures to support needed policies or to follow public health recommendations.

Although this strategy could be helpful in minimizing some of the possible harms resulting from NID, these benefits would have to be weighed against increasing some of the negative epistemic effects. This strategy necessitates that scientists use time and resources to respond to dissenting voices, and thus it would exacerbate some of the problems that the ability to identify NID is attempting to address. At a minimum, targeting dissent for criticism is unlikely to reduce this problem, which as we have seen can have detrimental epistemic consequences.[13] Furthermore, it is not completely clear that this strategy would be very successful in limiting confusion, incorrect beliefs, and negative effects on public policy decisions. If scientists have to engage with the dissent so as to point out the various ways in which the research is problematic, this would in fact contribute to the dissemination of the dissenting views. True, such dissemination need not result in negative effects if it is clear to the public that the criticism aims at informing the public and policymakers about why the dissent in question is problematic. Two concerns, however, raise doubts about the potential effectiveness of this strategy. Given that scientists are in the business of carefully scrutinizing *all* research, often they would be directing appropriate criticism to some dissent that though incorrect is epistemically valuable in various ways. As a result, it might be difficult for many stakeholders to make fine distinctions about whether dissent is being criticized for being unsound or for being normatively inappropriate when they lack scientific expertise. If so, then, engagement with the dissent, so as to criticize it, might also fail to reduce the social problems that result from NID. Furthermore, if the example of climate change is any indication, challenging NID might not be particularly helpful. Criticism of global warming skepticism has often focused on pointing out the ways in which the research falls short of meeting scientific standards.[14] Granted, these criticisms are produced in a context where no agreement exists about criteria for NID. Still, the fact that intense criticism of climate change dissent has failed to eradicate public confusion about the strength of the evidence or the need to support certain energy policies calls into question the effectiveness of this strategy, even in a context where criteria to reliably identify NID exist.

Limiting Duties to Dissenters

In chapter 2, we noted that the valuable epistemic and social roles of dissent give rise to obligations within scientific communities to actively seek and engage dissenters. Indeed, presumably it is at least in part the discharging of these obligations that contributes to the epistemic and social problems mentioned earlier when scientists engage dissent that is in fact NID.

The reliable identification of NID could be helpful in imposing limits on such obligations. If the dissent in question fails to promote knowledge production and can actually hinder it, scientific communities have no obligations to seek and engage this dissent.[15] Scientists could therefore ignore NID in various ways. For example, scientific communities could exclude dissenters as reviewers of manuscripts or grant applications, as well as reject dissenting research for publication or publish it only with caveats that categorize it as a waste of time.[16] Aware that there is something objectionable about this dissent, readers could be more attentive to and critical of the methods and reasoning used. Similarly, conference organizers would be justified in rejecting dissenting research deemed normatively inappropriate. To the extent that the dissenting research is published or otherwise disseminated, scientists could justifiably avoid responding to it and engaging with the dissenters.

Presumably, limiting scientists' obligations to seek and respond to NID would help prevent at least some of its negative effects. Scientists' time and resources would not be wasted responding to criticisms that ex hypothesi are unlikely to advance knowledge and could even hinder it, and they would be free to dedicate their time to more productive epistemic tasks rather than engaging in fruitless debates. Moreover, excluding problematic dissenters from activities such as peer review or conference participation would likely make it more difficult for them to disseminate their views, at least within the scientific community. If scientists refused to engage with dissent that hinders knowledge production, or declined to treat dissenters as intellectual equals, such dissenting views might also receive less attention from the media or the public. If this were the case, then NID might be less likely to generate doubt or confusion among the public and policymakers.

Yet, even if the reliable identification of NID serves to impose limits on scientists' obligations to seek and engage such dissent, this is unlikely to prevent either its production or its dissemination. Dissenters would still be free to conduct dissenting research as long as they could obtain funding from a variety of sources. As just discussed, much of the funding for dissenting research regarding climate change or tobacco and alcohol safety, for instance, comes from private industry and think tanks.

Dissenters would also be able to publicize their views, create journals and think tanks, organize conferences, and disseminate their views to the public and policymakers in ways that could contribute to confusion, the stalling of needed public policies, and the thwarting of important public health measures. Indeed, this seems to be what some climate skeptics are in fact doing. Because they believe that the mainstream scientific community is dismissing their work, they have begun to create their own journals, where those more sympathetic can review dissenting research on climate change. For instance, some have dubbed the journal *Energy & Environment* as the "journal of choice for climate skeptics" because it publishes more articles interpreted as supporting skepticism about anthropogenic climate change than any other journal.[17] Critics charge that this journal is not genuinely peer reviewed in the traditional sense but, rather, a thinly veiled attempt to bring credibility to climate skepticism in order to advance a particular political agenda.[18] Nonetheless, the work is being published and disseminated. Climate change dissenters have also been quite successful in publicizing their views through popular books, blogs, and the media.[19] Similarly, vaccine dissenters have eschewed scientific publications in favor of popular books, Internet sites, public rallies, and television talk shows and seem geared toward the public.[20] Given dissenters' ability to use a variety of resources to publicize their views, using criteria for NID to impose limits on the obligations of scientific communities to engage dissent will not be particularly helpful in preventing widespread dissemination of such views and hence, in preventing the problems that result from such dissemination.

Moreover, because the scientific community's disregard of their views is unlikely to deter these dissenters, failing to engage with NID might actually exacerbate some of the epistemic and social damages. If scientists fail to respond to the problematic dissent, the public and policymakers may see it as more plausible or weighty than it really is. This is so because scientists are uniquely situated to explain why, for example, some methodology or assumption is incorrect or problematic, given the available science. They are usually best able to interpret scientific evidence and to judge its strength. Without the engagement of the scientific community, then, dissenting views could well gain authority rather than disappear. This could aggravate problems of public confusion and stalled policies. Furthermore, in a context where powerful political and financial interests are at stake, and where dissenting views can flourish unchallenged, intimidation of scientists, with its attendant adverse epistemic consequences, is unlikely to disappear and arguably could increase.

In addition, a failure to seek and engage some dissent on the grounds that it is normatively inappropriate can generate suspicion among members of the public or policymakers that scientists are being dogmatic or are overlooking their own fallibility. This is what happened in the highly publicized case of hacked emails from a computer server at the University of East Anglia's Climate Research Unit (CRU).[21] Climate change scientists at the CRU came under fire because in some of those stolen emails, researchers were apparently boasting about how they could prevent articles with even the appearance of questioning anthropogenic climate change from surviving peer review.[22] The Inspector

General later found that there was no evidence that the CRU scientists had abused their authority in the peer-review process and that apparent attempts to block certain papers from being published were typical of the sort of robust debates over content that can occur in peer review.[23] Even though the scientists in question had good reasons to disregard research by climate skeptics, the attempts to exclude such dissent and the failure to engage it left the impression that climate scientists had something to hide.[24]

Granted, perhaps the public's reaction results from the fact that disputes persist—whether reasonably or not—about the epistemic value of climate change skepticism. If criteria for reliably identifying NID were available, that could minimize the possible public suspicion. This seems plausible. Presumably the public could be informed about the existence of such criteria and about their role in reliably identifying dissent that is epistemically value-less and might be detrimental. Equipped with such information people would be aware that the scientific community has legitimate reasons for failing to engage with some dissent. Nonetheless, it might not be easy for laypeople to make judgments about whether the criteria in question apply or not to a particular case of dissent. After all, much dissent—perhaps most—would still be epistemically and socially valuable—even if ultimately rejected, and thus scientific disputes in areas of relevance to public policy would still occupy a significant amount of attention.

Guiding Public Beliefs

Given the socially adverse consequences of NID, perhaps the identification of this type of dissent should be directed toward laypersons rather than scientists. That is, policymakers, the media, and the public could use the criteria in question to identify dissent that they should disregard and to understand why the scientific community's decision to ignore it is legitimate. Indeed, some have aimed to do just that by providing laypersons with criteria for determining when a scientific controversy has been "artificially created" so that they will discount such dissent and trust rather than doubt the existing scientific consensus on a particular issue.[25] Although it is true that using the criteria for this purpose would not help prevent the existence of NID, it would likely decrease the incentive to disseminate these dissenting views. After all, one important reason why using resources to widely publicize NID can be worthwhile is that its production results in confusion about whether a consensus view exists, and such misperceptions can be useful when trying to deter particular policies or advance others. But if the public can make use of some criterion that would allow them to determine whether some dissent is normatively inappropriate, then that would decrease the incentives to disseminate it. Minimizing the diffusion of NID would plausibly limit confusion and its effect on policy decisions.

Whether this strategy is likely to work, and thus whether it would be able to prevent or limit the negative impacts of NID, inevitably depends on the criteria to identify such dissent. If the criteria were relatively straightforward (such as the existence of financial conflicts), then laypeople could presumably make correct judgments on their own about what dissent they should disregard, and this would be quite helpful in preventing confusion or stalled regulatory policies that stem from such confusion. Journals, conference organizers, and even the media could all require disclosure of conflicts of interest, for instance. But, as our evaluation of various proposed criteria shows, finding a simple criterion that laypeople could use confidently is unlikely. Arguably, identification of NID will involve the ability to assess the scientific content— for example, the interpretation of the data, methodological issues, the strength of the evidence, and so on—in one way or another. But the general public and policymakers lack the necessary expertise to be able to do so, and thus they will be unlikely to use the criteria adequately. If so, determining whether the criteria in question apply to a particular instance of dissent will require the expertise of scientists. The involvement of the scientific community could presumably help laypeople make use of the criteria in more consistent ways, but it would also limit the value of this strategy, as scientists would still need to use time and resources to at least explain why they believe they are justified in classifying some dissent as problematic. How successful these discussions are would also affect the degree to which the social damage of the dissent could be minimized.

Emphasizing Consensus

Identifying NID could also be useful as a way to single out cases where scientists and science scholars have special obligations in communicating with the public. When facing cases of NID, scientists could do more to publicize the existence of a consensus and use more care when they communicate internal disagreements so as to present a united front to the public. Some argue that when the preponderance of evidence is clear, rather than being excessively careful about the existence of uncertainty of any scientific knowledge claim, scientists should be forceful about the existence of a consensus.[26] The presence of a scientific consensus is presumably a marker of objectivity and a good proxy for reliable knowledge, and thus it provides good reasons for the public to discount dissenting views, accept the consensus claims, and support related policies.[27] This might help inoculate the public against attempts to generate doubt about the state of scientific knowledge and prevent the stalling of necessary policies.

Some scientists and science studies scholars have pursued this strategy in response to climate skeptics, and have made an effort to empirically demonstrate and publicize the existence of a scientific consensus on anthropo-genic climate change.[28] Some scholars have thus focused on emphasizing that analyses of peer-reviewed papers on climate change, signed public statements by climate scientists, and surveys of members of scientific professional organ-izations have all shown that 90 to 98% of scientists publishing research on cli-mate change agree that anthropogenic climate change is occurring and likely to have significant impacts.[29]

Recent empirical studies suggest that this strategy is promising and sometimes successful in addressing doubt and confusion among the public and policymakers created by climate change dissent.[30] Evidence also suggests that the mistaken belief that there is no current scientific agreement among experts regarding anthropogenic climate change is strongly associated with reduced levels of support for mitigation policies.[31] Similarly, studies show that people were more likely to believe in anthropogenic global warming when the existence of a scientific consensus was emphasized and that belief in such a consensus was a gateway belief to increasing public confidence that climate change is happening, that human activities are causing it, and that it poses a worrisome threat.[32] In turn, the presence of these beliefs were predictive of support for mitigation policies.[33] Presumably, demonstrating the strength and breadth of the existing scientific consensus helps put dissenting views into perspective, so that they are not given more weight than they deserve. Similar evidence also indicates the importance of emphasizing the medical consensus regarding childhood vaccine safety.[34] Those studies suggest that when the public is clearly informed of the existence of an extensive agreement among medical professionals regarding the safety of childhood vaccines, people have fewer concerns about health problems, are less likely to believe in the vaccine–autism-link, and more likely to express public support for vaccines.[35]

If this evidence is correct, highlighting the existence of scientific con-sensus when in the presence of NID might help minimize negative social effects such as public confusion, stalled policies, or noncompliance with health regulations resulting from such dissent. It is not clear, however, how effective this strategy really is. Some data show no impact on people's willing-ness to act in certain ways or on their policy preferences as a result of pro-viding information about the scientific consensus.[36] Moreover, emphasizing the existence of a consensus among climate change scientists is not a new phe-nomenon.[37] Nonetheless, opinion polls show that people's beliefs about the existence of a scientific consensus have remained relatively stable.[38] It might well be that, even if in a laboratory context people are persuaded by claims about the very high number of scientists who agree on the existence of an-thropogenic climate change, many other influences present in the real world counter the force of consensus messages.[39] In the real world, those factors seem to apply to people's assessment of evidence regarding the existence of scientific consensus, just as they do to their evaluation of other claims related to climate change.[40]

Whether or not this strategy is effective, emphasizing consensus as a response to NID raises some serious concerns. Sometimes, the emphasis on a consensus is achieved by what some refer to as "masking disagreement" by "joint acceptance."[41] In order to present a united front to the public and establish that a consensus exists, scientists may publicize only those scientific claims about which they can all agree, while omitting or downplaying those about which disagreements exist. For instance, as we have indicated, although there is widespread agreement that average global temperature is increasing, general disagreements also exist about how much warming will occur, how quickly it will occur, and how best to prevent or mitigate it. Because presenting these disagreements to the public is likely to create uncertainty not just about the particular aspects about which there is disagreement but also about those for which agreement exists, scientists might conceal discrepancies by presenting only very general claims for which there are a consensus.[42]

Minimizing disagreement can also be the result of drawing attention to broad and simple consensus claims while failing to place this emphasis within a context that makes clear the existence of divergences about other claims. Studies highlighting the consensus on anthropogenic global warming often suffer from this problem.[43] They focus on a very general claim, such as the claim that anthropogenic global warming exists and is likely to have serious consequences. But agreement about this general claim need not involve accord regarding how much warming will occur, or how quickly, or what the specific impacts will be in particular regions. Indeed, some climate scientists believe that current models *underestimate* the impacts of average global warming, but they are nonetheless counted as members of the consensus view, broadly described.[44]

Emphasizing the existence of a consensus can thus result in various problems. Insofar as calling attention to the consensus is performed at the expense of playing down important disagreements, this can obscure information that is necessary for well-grounded policy.[45] For example, as mentioned earlier, there are disagreements about how much global warming is likely to occur, the extent to which factors such as aerosols or cloud formation might amplify warming effects, how quickly warming will occur, and what specific impacts will be for particular regions. In trying to minimize disagreements about the amount of aerosols or cloud feedback in global warming, climate models might be underestimating their effects on rising temperatures. Such underestimation could result in polices that prioritize efforts to mitigate greenhouse gasses over efforts to adapt to rising sea levels and extreme weather events.

Likewise, emphasis on expert agreement regarding the safety of childhood vaccinations can result in policies that fail to implement appropriate systems to monitor the safety and efficacy of approved vaccines, to incentivize the incorporation of new scientific evidence into vaccine design, or to promote the development of compensation programs that do not provide adequate remedies for vaccine-related injuries.[46]

The desire to emphasize consensus and present a united front to the public may also result in self-censorship or a "chilling effect" that can have negative epistemic effects.[47] Scientists may be reluctant to openly consider any scientific claim that could be interpreted as undermining the consensus, for fear that they will be accused of being denialists, faulted for contributing to confusing the public and policymakers, censured for abating deniers, or criticized for undermining needed public policy.[48] For instance, as we said before, there are genuine disagreements about how to best model for cloud formation, water vapor feedback, and aerosols in general circulation models (GCMs), all of which have significant impacts on the magnitude of the warming likely to occur.[49] Some argue that scientists are self-censoring so as not to contribute to public confusion and that, as a consequence, current climate models are actually far more conservative than is warranted.[50] Similarly, because of various concerns many parents have regarding childhood vaccinations, some of them request alternative vaccination schedules that increase the time between vaccinations or that reduce the number of vaccinations in a single well-child visit.[51] The safety and effectiveness of alternative vaccination schedules is, however, unknown. Nonetheless, pediatricians and public health advocates are reluctant to openly question the traditional schedule or to call attention to the fact that different countries have different childhood vaccine schedules, for fear that the public will misinterpret this information and undermine public confidence in the safety of infant immunizations.[52] But in doing so, they are limiting the possibility of conducting research on alternative immunization schedules and thus of gathering evidence about their safety and efficacy.[53]

Thus, while the reliable identification of NID could be useful in helping scientists and science scholars determine when emphasizing the consensus and forcefully communicating expert agreement is necessary, there are drawbacks to this strategy. On the one hand, making the public and policymakers aware that a consensus exists can limit confusion about the state of the science and make people more likely to support policies and behavior consistent with the evidence, addressing the negative social effects of NID. On the other hand, an emphasis on the consensus can come at the expense of masking complexities regarding the scientific evidence and concealing important disagreements that can be relevant for policymaking. Thus, this strategy can have adverse social effects. Similarly, insofar as the ones working to emphasize consensus views are science scholars and journalists, this strategy would limit epistemic damages resulting from NID by allowing scientists working on the relevant areas to dedicate their time and resources to research rather than to respond to criticisms. However, this strategy fails to address the epistemic concern that NID can lead scientists to make scientific claims that are more conservative than the evidence warrants. Although the reasons for such conservatism are different—intimidation when NID is rampant; a desire to present a united front when emphasizing the consensus—in both cases, scientists might be compelled to understate their claims. Moreover, a focus on the consensus can lead scientists to decide against conducting research on certain contested areas for fear that doing so will increase confusion, distrust, or opposition to needed policies. If so, then this strategy would also fail to address concerns about epistemically detrimental effects of NID.

Discrediting Dissenters

Identifying NID could also be useful as a way to determine when scientists and others should discredit dissenters. Scientists, and particularly science studies scholars, often scrutinize and call attention to problems with the science that grounds dissent they believe is problematic. They have also focused on criticizing the dissenters, both those who produce the research and those who disseminate it.[54] In some cases, scholars have revealed the dissenters' credentials in order to demonstrate that they lack relevant expertise.[55] In other cases, they expose the dissenters' financial or political ties to think tanks or private industry.[56] Still in others, scientists have attempted to discredit certain kinds of dissent because they take such dissent to be motivated by a particular political agenda, even when no formal financial or organizational ties exist.[57]

Exposing features of dissenters that might make them unreliable, such as the presence of financial conflicts of interests or ideological ties that could lead to confirmation bias, can certainly be important and useful for aiding the public in evaluating what information is reliable. Those with more power tend to have more financial and political resources for disseminating dissent. This can make certain cases of dissent seem to have more weight or be better supported by the evidence than is the case. Revealing financial or ideological conflicts of interest could lead members of the public and policymakers to scrutinize the dissenting views more carefully and to be less inclined to believe NID. This could ultimately minimize confusion or stalled needed policies. Moreover, insofar as dissenters are concerned with their reputations and their careers, this strategy could minimize not just the consequences of NID but also its production.

However, discrediting dissenters is not free of problems. It could discourage scientists from engaging in patenting activities, as well as in industry–academia collaborations, given that these practices involve conflicts of interest. Insofar as patents and collaborations between industry and academia are deemed important to advance research and provide the public with new medical interventions,[58] this strategy could actually result in epistemic and social damages.

Moreover, this strategy could dissuade scientists from conducting legitimate studies or publishing findings that run counter to dominant views.[59] That is, scientists may fear that the risks of dissent are too great. After all, as mentioned earlier, even if criteria to identify reliable dissent exist, their application would still require judgments about whether a particular instance of dissent meets the relevant criteria. It would thus not be unreasonable for scientists who have conflicts of interest, for instance, to be concerned about conducting dissenting research or expressing dissenting views. Indeed, there is already evidence of a bias toward conservatism in publishing and the awarding of grants and fellowships.[60] Such epistemic conservatism can deter other scientists from research that is highly innovative or that clashes with widely accepted theories because doing so could harm their careers. If so, then whatever epistemic and social benefits can result from this strategy could be at the cost of constraining the valuable role that legitimate dissent has in science.

A further, related worry is that discrediting dissenters can encourage them to do the same to consensus-view scientists. Financial ties and ideological motivations can also be present in the research done to support the scientific consensus. Legitimizing attacks on a person, rather than exclusively on the epistemic merits or flaws of the individual's research, sends the message that this is an appropriate type of response against those with whom we disagree. Indeed, as we saw earlier, climate and vaccine skeptics are quick to point out financial conflicts of interest that exist among climate scientists or spokespersons for vaccine safety. If the public believes financial or ideological ties are a legitimate reason to disregard certain research or views, this may give dissenters an incentive to expose the political and economic ties of scientists who support the consensus. At a minimum, this strategy is unlikely to reduce the intimidation of scientists, and it is likely to exacerbate it, thus hindering scientific progress.

Conclusion

Although finding criteria to identify NID might appear a promising strategy to limit the negative epistemic and social impacts that such dissent can have, we have shown here that the promise is likely to go unfulfilled. Using such criteria to determine when to prohibit dissent, limit engagement with dissenters, emphasize consensus, discredit dissenters, guide public beliefs, or target dissent for particularly rigorous criticism could be helpful in trying to address some of the damages that can result from NID. Nonetheless, adopting any of these strategies, or even a combination of them, seems to either be ineffective or create more problems than it solves. We have argued that some of these strategies could be helpful, but are either impractical or implausible, such as prohibiting NID in democratic societies. In some cases, as with emphasizing consensus, these strategies could help address some of the problems, but at the expense of creating other negative consequences. Some of the strategies risk exacerbating the very problems they are intended to address, as is the case of imposing limits to scientists' epistemic obligations to engage dissenting views.

To be sure, these strategies could be helpful or even necessary to achieve goals others than to prevent or address the problems that can result from NID. For instance, discrediting dissenters—or consensus proponents, for that matter—who have conflicts of interest or lack appropriate expertise, or revealing features of the dissent that make it dubious, such as questionable evidence or suspect methodological practices, can provide people with relevant information in order to assess the reliability of the research. Similarly, insofar as a consensus exists, pointing out its existence also supplies people with appropriate information they might need when making decisions about what policies might be better grounded.

Nonetheless, given the difficulties in finding criteria to identify NID and considering the fact that the ability to pinpoint such dissent is unlikely to help prevent or reduce its adverse social and epistemic effects, it is thus worth examining whether there might be other ways of addressing the epistemic and social problems of concern that do not depend on reliably identifying NID. To that task we devote the next part of the book.

7

The Relevance of Trust

A recent study by the Pew Research Center shows that although U.S. scientists and the public share broadly similar views about the overall place of science in America, they have different beliefs about various scientific issues.[1] For instance, 88% of American Association for the Advancement of Science (AAAS) scientists consider genetically modified (GM) foods to be safe for consumption, but only 37% of the public say such foods are safe and a majority of them (57%) think GM foods are unsafe to eat.[2] Similarly, while 65% of the public say that humans have evolved over time, 98% of scientists believe such is the case.[3] And although 87% of scientists say climate change is occurring due to human activity, only 50% of the public agree with such a statement.[4] When asked about their beliefs regarding the degree of scientific consensus or understanding on some of these topics, 29% of the public say that scientists do not agree about human evolution[5] and 37% believe that scientists generally disagree that the earth is warming because of human activity.[6]

Although the statistics vary, the gap between scientists' and the public's beliefs regarding various scientific claims or the existence of a consensus on some research areas is a worldwide phenomenon. For example, 27% of Western Europeans are not comfortable with the risks associated with GM crops;[7] 22% of Canadians disagree that climate change is largely the result of human activity;[8] and a majority of the public (56%) in the Dominican Republic do not believe that humans have evolved.[9] Similarly, only 11% of the public in the United Kingdom say they are aware that nearly all scientists agree climate change is occurring due to human activity.[10] Of course, this gap is relevant not only for epistemic reasons. As we have seen, failure to believe certain scientific claims can also have important effects on people's behaviors and their support for public policies consistent with the evidence.[11]

Why this gap? Doubtless, a variety of complex factors affect these general differences between the public and the scientific community, including educational attainment, political affiliation, social identity, and religious beliefs.[12] People's access to and familiarity with reliable sources of scientific information can clearly contribute to this gap. Various cognitive biases are also likely to play a role. Individuals often are more inclined to accept claims that are consistent with their other existing beliefs, or that they take to support particular favored policies.[13] In such cases, individuals may inappropriately attribute more weight to such claims than to those that would challenge their accepted positions. Nonetheless, as we have seen, many scholars argue that the existence of problematic or normatively inappropriate dissent (NID) in these areas is also responsible for the misinformation and confusion affecting the publics' beliefs.[14] Problems such as stalled public policies, refusal to accept public health recommendations, and wasted scientific resources are thus attributed to the confusion that NID creates. While we agree that this type of dissent can indeed have these effects, we argue that a focus on NID as the culprit of the various social and epistemic problems that we have discussed is unlikely to help us prevent or minimize these problems. In previous chapters we argued that reliably identifying NID is a task easier said than done. We evaluated various criteria that could be used for such a task and concluded that none of them is successful. Even if some such successful criteria could be proposed, we argued that nonetheless the ability to reliably identify NID is unlikely to be useful in preventing or reducing the problems that can result from such dissent. As we have shown, attempts to preclude these types of dissenting views are likely to fail and can actually undermine epistemically valuable dissent.

But if finding criteria to reliably identify NID is unlikely to be effective or useful, then what should be done about this dissent and its adverse consequences? In this and subsequent chapters we seek to call attention to other factors that arguably contribute to making NID particularly damaging. We focus on two such factors: a lack of warranted trust in the scientific community, and an excessive reliance on scientific information when it comes to assessing policy decisions. We contend that a context where the trustworthiness of scientists is called into question allows NID to take hold and erroneously affect people's beliefs and ultimately their actions. If this is so, attending to factors that affect people's trust in the scientific community can provide resources for limiting or minimizing the epistemic and social damage that NID can inflict. Yet despite the importance of trust for people to rely on scientific claims that often depend on expertise, more is needed to address public resistance to particular policies or actions. Value judgments are essential to such decisions, and in some cases that is where the real disagreements lie. Insofar as this is so, a focus on dissent over the facts neglects the real debate and allows parties to talk past one another.

This chapter offers a brief overview of the importance of epistemic trust and the relevance that scientific institutions and practices have in promoting or undermining warranted trust. We also make a case for why a context where trust in scientific communities is precarious makes NID more likely to create damage than it otherwise would. Chapters 8 and 9 will examine various social and institutional factors, as well as concerns about the negative influence of nonepistemic values in scientific knowledge production that cast doubts on the trustworthiness of scientists and thus undermine public trust in scientific communities. In chapter 10, we address the excessive focus given to scientific knowledge in policy disputes and argue that it diverts attention away from the values that necessarily play a role in policymaking.

The Relevance of Epistemic Trust

That trust, moral and epistemic, is central to the doing of science is uncontroversial. Whether they do so completely or not,[15] scientists must place trust in the testimony of colleagues, their techniques, experiments, data, results, and theories in order to be able to conduct research.[16] They have to put trust in the institutions that govern scientific conduct. Scientific projects often involve teams of researchers from multiple disciplines, working at various institutions and in different countries. Hence, researchers are epistemically dependent on one another, making trust all the more important.[17]

But epistemic trust is also significant to the multitude of interactions between science and society.[18] Scientific knowledge is increasingly complex, abstract, and reliant on intricate technological devices and cognitive tools such as models. Making sense of scientific phenomena such as molecular biology, neuroscience, atoms, chemical reactions, and so on requires a significant amount of expertise. Thus, to understand particular phenomena, members of the public must therefore trust scientific experts and rely on the information they provide. Yet, they must also do so in order to participate in democratic discussions that implicate scientific knowledge, such as whether particular toxins are safe, whether a drug is beneficial, what the impacts of nuclear waste might be, what the benefits and risks of stem cell research are, how global warming can affect food production, and so on. Lack of warranted trust on the side of the public regarding scientific testimony can thus be an obstacle to fully realizing science's goal of benefiting society.

Although the relevance of trust in science is not disputed, what constitutes trust is a matter of debate.[19] It is clear, however, that trust is a complex, multifactorial phenomenon. In general, one trusts others to do something when one relies on them to do it and to do so for the right reasons. Trust, therefore, need not be complete, as in A trusts B. Indeed, it often is not. Usually one trust others to do something in particular. In the case of scientists, we trust them to provide us with reliable and relevant knowledge. Given the complexity of scientific knowledge and its connection to expertise, our trust in scientists is specific to their disciplines. We trust geneticists but usually not geologists to produce reliable knowledge regarding genetic inheritance, for instance.

When we trust, we are vulnerable to others. Hence, trust is risky; our trust can be betrayed. If people trust scientific experts to produce and disseminate sound knowledge and scientists fail to do so, people will have incorrect beliefs and make inadequate decisions. Given the risks involved in trusting, we want to place our trust only in those who are trustworthy. What trustworthiness involves, however, is also a matter of philosophical controversy.[20] Generally, experts are trustworthy when they are at least honest, competent, benevolent, and reliable. Empirical research also shows that people judge scientific experts as trustworthy when scientists are competent, have moral integrity, and are concerned with applying their work for the benefit of society.[21]

But assessing whether someone is trustworthy is not always an easy task. Prior experiences with trust are likely to influence such evaluations, as will social cues that can be more or less dependable. Unsurprisingly, we sometimes make incorrect judgments about whether someone is trustworthy and thus we misplace our trust, trusting those who are undeserving of our trust. But, we can also fail to trust those who are in fact trustworthy and who should therefore be trusted. This latter type of misjudgment results in a credibility deficit.[22] Sometimes the misjudgment is the result of innocent errors, but as feminist philosophers in particular have argued, it often can result from prejudice.[23] Trustworthiness is thus a property that belongs to the trustee, while credibility constitutes the truster's perception of it.[24] Hence, for science to provide public benefits, for it to achieve its practical goals, scientists must be not only trustworthy but also credible, for otherwise the public is unlikely to believe their testimony.[25]

Clearly, some social and institutional practices regarding science and scientific testimony will be more or less likely to promote the character traits and dispositions that make scientists trustworthy and credible. Different ways of organizing the training of scientists, the assessment of their contributions, the mechanisms of research funding, or the reporting and dissemination of scientific testimony—all can have significant effects on scientists' ability to actively and positively engage with the fact that others are dependent on their research and their testimony. But scientific institutions and practices can also affect how laypeople perceive scientists' competence, honesty, or benevolence. The best systems, then, to promote and sustain warranted public trust in science will be those that both contribute to the cultivation of scientists' trustworthiness and that reliably indicate to others such trustworthiness so that scientists can also be credible.

In chapters 8 and 9, we discuss several factors that are not conducive to cultivating trustworthiness or that undermine credibility or both. Chapter 8 focuses on various aspects related to the current context of scientific production. We consider the role of the increasing commercialization of science, with its effect on the common good and on conflicts of interests, and of scientific misconduct in eroding public trust. Chapter 9 discusses another factor that can contribute to casting doubt on the trustworthiness of scientists: the belief that the results of controversial areas of inquiry, such as climate change, vaccine safety, and GMP research, are biased because they are influenced by scientists' values. We explore how beliefs about the role of nonepistemic or contextual values in scientific inquiry can lead people to question the trustworthiness of scientists and can undermine trust in scientific claims. Correcting and improving these various factors is important in preventing damage to warranted trust in scientific communities. Equally important for our purposes, such actions can provide resources in dealing with the challenges that NID can create. Although, as we will see, correcting these factors might be difficult, we consider some possible—but necessarily incomplete—strategies for changing these trust-undermining practices so as to better facilitate and maintain warranted public trust in scientific communities. We also indicate where further research is needed to identify more effective mechanisms for establishing and maintaining warranted epistemic trust.

Lack of Trust and NID

No one would deny that epistemic trust is necessary in conducting research and in ensuring that science achieves its epistemic and social goals. But what is the relevance of trust, its presence or absence, to the epistemic and social problems that can arise from instances of NID? As indicated at the beginning of this chapter, evidence shows a significant gap regarding some scientific claims between scientific experts' beliefs and those of the lay public. Although, as we said, many factors are likely to contribute to this gap, many scholars argue that problematic dissent is a major culprit.[26] NID can produce confusion and that confusion can also lead people to oppose needed public policies or ignore expert advice.

Recall the case from chapter 1 of former South African president Thabo Mbeki, who firmly opposed government support for existent HIV antiretroviral (ARV) drugs and sought to limit their use in the country.[27] Mbeki's government withdrew support from clinics that had begun using ARVs, obstructed Global Fund grants, and restricted use of ARVs.[28] Researchers have estimated that these restrictions during the Mbeki presidency resulted in at least 330,000 South Africans' premature death and the birth of 35,000 HIV-infected babies.[29] Evidence shows that in justifying these policies, Mbeki, and his minister of health, Manto Tshabalala-Msimang, appealed to dissenting views of Peter Duesberg and colleagues who contended that HIV did not cause AIDS and that ARVs were not useful against the disease.[30]

Why were Mbeki and other government officials receptive to Duesberg's dissenting views despite the existence of overwhelming evidence and consensus in the global scientific community that HIV caused AIDS and that ARVs were the most effective way to control HIV infection? Ignorance, various psychological factors, and character flaws could explain their inclination to believe the dissenting views. But evidence indicates that Africans' experiences with the scientific community had not been conducive to promoting and sustaining warranted trust. AIDS emerged in the 1980s when the struggle against apartheid, led by the African National Congress (ANC), had intensified. Members of the ANC and black South Africans were targets of increasing military aggression and covert chemical and biological warfare programs aimed at eliminating ANC leaders and causing widespread infertility among blacks.[31] This context, then, created a belief system and set of experiences that shaped the ways in which Mbeki's government interpreted mainstream scientific views about HIV and AIDS.[32] Mbeki himself characterized mainstream scientific views about HIV as part of the capitalist–apartheid ideology against which ANC had fought. An ideology, he contended, that was designed to make black South Africans believe in their own inferiority by reminding them, among other things, of their role as "germ carriers."[33] Moreover, scientists themselves often exacerbated this perception by characterizing the severity of African AIDS epidemic in ways that reinforced racist beliefs about the sexual propensities and promiscuity of Africans.[34] In addition, an early attempt at testing AIDS vaccines in Africa, which although justified by certain scientific considerations, raised ethical concerns about the use of Africans as guinea pigs for clinical trials.[35] There were also significant worries that Western pharmaceutical companies were making alarming announcements regarding HIV and AIDS in order to maximize their profits.[36] When Mbeki was elected in 1999, South Africa faced an economic crisis and ongoing clashes with drug companies over South Africa's right to parallel importation of pharmaceuticals, which would have significantly lowered their costs.[37]

The work of a few dissenting scientists such as Duesberg clearly enabled AIDS denialism in Mbeki's government. But the dissent seemed more compelling and credible to many South Africans because of an existing lack of trust in the scientific community, grounded in its support for racist views about Africans and blacks, its failure to achieve cultural competency while conducting research in Africa, and its engagement in unethical research practices with African subjects.

We agree, then, that NID can have adverse consequences for public policies, individual actions, and for science. Nonetheless, we believe that a context in which the trustworthiness of scientists is called into question, where trust in scientific communities is precarious, allows NID to be more damaging than it could otherwise be. That is, NID takes advantage of an ongoing crisis in credibility. When the public fails to believe that scientific communities deserve their trust, when it doubts the trustworthiness of scientists, questions about the reliability or soundness of their testimony will find fertile soil.

Lack of trust can thus exacerbate the negative impacts of NID in several ways. First, in a context in which people are suspicious of the scientific community, the existence of dissenting views is more likely to lead to increased hesitation and uncertainty about who or what to believe and what actions might be justified. Whether or not the public is compelled to believe the dissenting view, people can take the presence of dissent as evidence that there is significant disagreement in the scientific community, and that no particular view is yet sufficiently reliable. This can explain at least some instances of climate skepticism. That is, people may be suspicious of both those who maintain and those who challenge the consensus view on climate change, and thus they are unsure about who or what to believe. Thus, it may seem plausible to them to simply suspend belief or action. Of course, inaction can also have disastrous consequences.

Second, lack of trust in scientific communities can enhance the plausibility of dissenting minority views. In a context where warranted trust in scientific communities is actively cultivated and sustained, the public is more likely to appropriately recognize dissenting views as minority ones that must be weighed against the strong evidence that an existent consensus provides. Thus, it would be more difficult for the public to believe dissenters or to be confused about the strength of the evidence or the degree of agreement between experts. This is so because we usually require a higher standard of evidence to call into question the trustworthiness of those whom we trust. Hence, in a context where appropriate trust is nourished, people are likely to require high evidentiary standards in order to give credence to dissenting views. That is, not only must dissenters produce some evidence that the consensus is false but also enough evidence to overcome the presumption of plausibility that a trusted consensus would provide. However, when scientists defending the consensus are seen as untrustworthy, or their trustworthiness is in question, criticisms of the consensus have more traction. The burden of proof thus shifts to those who maintain the consensus view to prove that it is reliable. Worse still, insofar as the public has misgivings about the trustworthiness of scientists, certain responses to the dissenters—for example, attempts to exclude or silence them, or accusations about their motives—can actually backfire and lead people to give more credit to the dissenter's claims, thus steering them to oppose policies or behaviors that a consensus would support.

Third, insofar as dissenting views involve criticisms of consensus scientists that cohere with background assumptions related to people's distrust, such dissenting views can have more influence than they otherwise would. We have seen how this occurred in the case of Mbeki's AIDS denialism: Duesberg's research reinforced Mebeki's beliefs about racist, imperialist, and capitalist Western science. A similar dynamic may also be present in some cases of climate change skepticism. Polls in the United States, for instance, have found that 69% of adults polled believe it was at least somewhat likely that climate scientists have falsified evidence for global warming[38] and 37% believe that global warming is a hoax.[39] A recent Pew Research Center report also shows that large numbers of Americans believe that scientists allow conflicting interests, such as career, political leanings, or their desires to help related industries, to influence climate research findings at least some of the time.[40] McIntyre and McKitrick's criticism of the "hockey-stick graph" contended that Mann and his colleagues—the authors of the graph—had cherry-picked proxy data and used questionable statistical methods.[41] Although these claims were subsequently dismissed by the scientific community,[42] many gave the dissent more attention and weight than it deserved because it was consistent with fears that climate scientists were untrustworthy. Similarly, those who challenge the safety of genetically modified products (GMPs) often point to the fact that most of the researchers who support their safety are funded by agribusinesses such as Monsanto that would benefit from wide-scale implementation of genetically modified crops.[43] Criticisms that such products are not safe, or that safety has not been proven, can thus be given more weight because it is consistent with people's suspicions that the consensus view may be unreliable. Those who are skeptical of the safety of pediatric vaccines are also worried that the pharmaceutical industry, with its less than stellar history of downplaying or ignoring drug risks, has funded much of the science supporting the safety of vaccines. People can thus give dissenting views about vaccine safety more weigh, even when evidence that, for example, there is a link between the MMR vaccine and autism is lacking.[44]

Conclusion

Epistemic trust is crucial to interactions between science and society. Laypersons must be able to trust scientific knowledge that is increasingly complex and necessitates highly specialized expertise. Trust in science is important not only for arriving at true beliefs but also for people's ability to effectively participate in democratic policy debates. In a context where the trustworthiness of scientists is in doubt, NID is likely to gain more traction and be more damaging. Such a context undermines the credibility of a scientific consensus, which will fail to function as a compelling reason to accept particular scientific claims in the face of dissent. Such context also makes NID seem more plausible than it otherwise would, particularly when the dissent in question coheres with background assumptions related to the source of distrust.

   If we are correct about the role that trust in the scientific community plays in shaping the credibility of scientific claims, attending to ways to facilitate warranted epistemic trust can provide us with resources to prevent or reduce the negative consequences of NID. Facilitating and sustaining well-grounded public trust in science can have the added benefit of improving relationships between scientists and society so as to ensure that the practical aims of science are achieved.

Before proposing strategies for facilitating warranted public trust in scientific communities, however, it is important to identify some of the significant institutional factors that contribute to calling into question the trustworthiness of scientists and to undermine their credibility. In the next two chapters, we aim to do just that. We focus on several current institutional factors that arguably undermine trust and thus have contributed to make the cases of dissent discussed in this book more troublesome. In chapter 8, we attend to the increasing commercialization of science, with its effects on the common good and on conflicts of interest, as well as the phenomenon of scientific misconduct. In chapter 9, we discuss concerns about the negative influence on nonepistemic or contextual values in science. In each case we propose some strategies to deal with these problems in ways that we believe contribute to promote warranted trust in scientific communities. Transforming current institutional practices in ways that are trust promoting will certainly be a difficult task, but if we are correct about the damaging effects that such trust-undermining practices have on scientific testimony, focusing on conducting research directed to address and solve these concerns is likely to be more successful in limiting the damages that NID can have. Moreover, insofar as factors that affect trust have other epistemic and social effects, focusing on them can have additional benefits that would be absent when focusing primarily on identifying and limiting NID.

8

Scientific Practices and the Erosion of Trust

In Henrik Ibsen's *An Enemy of the People*, the main character, Dr. Stockmann, confirms that the town's baths, which are viewed as essential to its economy, are polluted.[1] The pollution comes from tanneries and other industries that have contaminated the main source of the baths' water near Milldale and seeps into the baths' pump room. Dr. Stockmann assures everyone that the problem can be fixed by replacing the water system. For various reasons, however, people believe that he is exaggerating the dangers. They do not trust his testimony. They do not find him trustworthy.

Laypeople are dependent on the knowledge that scientists produce. Ideally, we should trust those on whom we are epistemically dependent when they are trustworthy and distrust them when they are not. If people fail to trust trustworthy scientists, their claims will not be credible when in fact they should be. If, on the other hand, people trust scientists when they ought not to, they risk believing their claims when they should not.[2] As we discussed in chapter 7, social contexts, institutions, and practices can be better or worse at promoting and sustaining warranted epistemic trust, a trust that is crucial not only to the production of scientific knowledge but also to its dissemination and use.

Feminist philosophers of science have called attention to the ways in which scientific practices and institutions have failed to sustain and have actively eroded the trust of particular communities.[3] Scientific communities have treated women and minority groups unjustly, science has been used to impose differential risks on these groups and maintain harmful gender and racial stereotypes, and research agendas have often ignored topics of importance to them. The sorry history of human-subject abuses in research has more often than not involved minority populations.[4] Similarly, the exclusion of women from clinical research with its consequent harms to women's health and well-being is well documented,[5] as is the use of gender stereotypes to ground scientific claims.[6] Likewise, scientific evidence is often used to justify decisions about the locations of high-level radioactive waste facilities, nuclear and other power plants, oil pipelines, or fracking in minority communities.[7] Unsurprisingly, and arguably reasonably, these populations tend to be suspicious of scientific communities and their claims.[8]

But some social and institutional factors related to the practice of science can affect the trust not only of particular populations but also of people in general.[9] In this chapter we explore some of those factors present in the current context of scientific production that arguably cast doubt on the trustworthiness of scientists and thus contribute to undermining the trust that people place in the scientific community and its claims. Specifically, we consider the role of the increasing commercialization of science, with its effect on the common good and on conflicts of interests, and of scientific misconduct in eroding public trust. We explain why this occurs and provide evidence to support the claim that trust has, in fact, been undermined as the result of these factors. We also consider strategies for changing these trust-undermining practices so as to better facilitate and maintain warranted public trust in scientific communities. If, as we have indicated in the prior chapter, lack of trust provides normatively inappropriate dissent (NID) with a more receptive environment and thus enhances the potential damaging effects that such dissent can have, then correcting those factors that undermine warranted trust in scientific communities can contribute to diminishing the negative consequences of NID.

Importantly, we do not claim that these are the only elements related to current institutional and scientific practices that can call into question the trustworthiness of scientific communities and undermine trust in scientists and their testimony. Indeed, many other factors are likely to do so. For instance, as the public is made more and more aware of the lack of reproducibility of many scientific studies,[10] this is likely to undermine people's trust in science. The discussion here is thus not intended to be an exhaustive exploration of the various factors that can negatively affect public trust in scientific communities. Trust, as we have indicated, is a complex phenomenon and various and heterogeneous elements can affect its presence and absence. We believe, however, that the factors we consider here are particularly salient regarding their harmful effect on people's trust. The increasing commercialization of science and of misconduct cases are common topics of discussion in the general and specialized media, and both affect multiple scientific disciplines. Furthermore, we do not claim that the effects these factors have on trust extend equally to all scientists or their testimony. In general, though, the aspects of current scientific practice we discuss here arguably cast suspicion on the trustworthiness of scientific communities and thus can negatively affect people's perception of it.

Commercialization of Academic Science

One of the factors that can call into question the trustworthiness of scientists and undermine warranted public trust in the scientific community relates to the increased commercialization of science in general and academic science in particular. The pressure to commercialize scientific innovations is pervasive and global and, not uncommonly, is presented as a social good that deserves institutional focus and support.[11] Private industry is a significant source of funding for research and development (R&D) worldwide. In the United States, for instance, industry funding accounted for 65% of total R&D spending in 2013, with federal funding accounting for 27%.[12] Private investments have become the largest source of funding in renewable energy research, in contrast to a decade ago.[13] Industry's financial support for biomedical research in 2012 accounted for 58% of total biomedical research funding in the United States.[14] In 2011, global biomedical research expenditures by industry were approximately $162 billion, while public funding was over $102 billion.[15] In food manufacturing and agricultural research, there has been a dramatic increase in private funding since 1950, and by 2010 it accounted for 45% of all agricultural research funding.[16] Although state and federal funding still constitutes the primary source of money for universities,[17] partnerships between academia and industry are more and more common and strongly encouraged.

The commercialization of academic science is particularly significant in areas such as biomedicine, pharmacology, genetic sciences, and artificial intelligence. In the United States, some areas critical to the public's interest, such as clinical trials, are now funded primarily by industry. A recent study shows that since 2006, the number of registered clinical trials funded by industry each year has grown by 43%.[18] But industry funding is not the only way in which commercial interests are entering the academic setting. Intellectual property rights, particularly in the form of patents and licensing agreements, are now common, as are academic start-ups. Indeed, academic patenting has been on the rise in the United States, with nearly 6,000 patents granted by the Patent and Trademark Office to U.S. academic institutions in 2014.[19] Pharmaceuticals constituted the largest technology category for U.S. academic patents in 2014, which made up 16% of patents to academic institutions. Biotechnology is now the second-largest category (13%) of university patents.[20]

In the United States, this state of affairs is the result of the Bayh-Dole Act of 1980. The act created a uniform federal patent policy that allowed universities to retain property rights to any patents resulting from federally funded research and to license these patents on an exclusive or nonexclusive basis.[21] It thus encouraged universities and individual researchers to patent and commercialize inventions that resulted from public funding. But the encouragement to patent and commercialize academic research is a global phenomenon,[22] and thus the effects of such commercialization on the trust placed in scientific communities is likely to be relevant worldwide, even if such effects vary due to particular political, cultural, and economic factors.

How can the increasing commercialization of science cast doubts on the trustworthiness of scientists? It can do so in at least two ways. First, it can undermine people's perceptions that scientific communities have as a primary goal to benefit the public with their research. Second, the existence of conflicts of interest can lead people to question the integrity of scientific research. In what follows we discuss how these factors influence trust in scientific communities and thus affect whether and to what extent the public will believe their claims.

UNRESPONSIVENESS TO THE COMMON GOOD

One of the primary goals of scientific inquiry is to generate justified truths. But as we mentioned in prior chapters, the epistemic merit of science is not limited to this goal. Producing *significant* knowledge is also an essential goal of scientific inquiry.[23] For instance, biomedical sciences do not simply provide us with facts about human biology; they also attempt to promote human health. Environmental sciences offer evidence about ecological niches, but they are also concerned with promoting ecological diversity. Engineering disciplines provide us with knowledge to construct devices that help us interact with and manipulate the world. Social sciences offer evidence about social interactions, group dynamics, and cognitive biases, but also allow us to develop strategies to promote or obstruct particular behaviors. Insofar as the use of scientific knowledge is a constitutive goal of inquiry, ensuring public trust is essential to fulfilling such goal. If the public considers scientists untrustworthy, it is unlikely to believe their claims, which will interfere with the practical goals of inquiry.[24]

Indeed, evidence shows that people's positive opinions about science stem at least in part from their beliefs about its practical goals. For instance, data from surveys show that the public in Europe and the United States holds a positive view of science in general, with a majority agreeing that scientific research makes life easier and has a positive influence in society and that scientists make a valuable contribution to society.[25] A majority of people in Europe and the United States also believe that scientists work for the good of humanity and that their work is directed toward making life better for the average person. These opinions are consistent with a view of science as having practical goals rather than simply epistemic ones.

How does the commercialization of science undermine the practical goals of science in ways that threaten public trust? Benevolence, as we saw in chapter 7, is one of the character traits associated with trustworthiness. In the context of science, such benevolence on the side of scientists is illustrated by scientists' concern with applying their work for the benefit of society.[26] The increasing commercialization of science, however, can have adverse effects on people's perception of scientists' benevolence and thus affect their credibility. Commercial interests can influence research direction or priorities toward the production of profits, regardless of public interests. Consider biomedical research, which has as a constitutive end to improve the health of the population. Commercial interests are leading research in directions that are arguably unresponsive to basic public health needs or that neglect the health needs of marginalized groups.[27] In the United States, for instance, evidence shows that public funding of medical research in 2010 was only marginally associated with disease burden.[28] In addition, despite the fact that health service innovations can reduce both morbidity and mortality and costs, only 0.3% of total health-care funding was directed toward such services, compared to 4% directed to new drugs and devices.[29] Globally, the situation is not very different. Noncommunicable or chronic diseases are now the leading global cause of death and disability, representing 63% of all annual deaths, but their funding is not consistent with their burden.[30] For instance, in 2011, $7.7 billion of global funds were allocated to HIV/AIDS projects, while noncommunicable diseases received only 1.5% of all health aid.

Multiple factors, other than disease burden, affect decisions about funding priorities, including scientific opportunities, the quality of existing research, and availability of infrastructure.[31] But commercial interests can surely affect the direction of research in ways that fail to attend to the needs of the public. First, because industry significantly funds biomedical research, it also has considerable power to determine the kinds of questions investigated,[32] as well as how problems and solutions are framed.[33] This can shape research programs and priorities in a variety of ways. In the case of biomedical research, because companies need products that can offset the expenses of drug and device research and development, they prefer strategies that minimize economic risks. This might include combining agents that are already approved or making slight chemical changes in order to extend patent rights on particular drugs, even though these changes in most cases bring little or no added benefit to patients.[34] Likewise, the food and beverage industries have funded chemists to engage in research aiming at making processed foods and sugary beverages taste better and more enticing, despite the fact that this may have disastrous public health consequences.[35]

Second, the development of products and services that can result in economic activity can be favored over less lucrative strategies, such as promoting lifestyle changes, ensuring access to adequate nutrition, developing strategies to reduce pollution, and improving working conditions, even when such strategies could be more effective in improving public health.[36] In the biomedical sciences, for instance, drugs, devices, diagnostic or predictive tests, and similar interventions result in financial benefits to the firms that invests in them, while service innovations and preventive measures, even when they can reduce morbidity and mortality, offer negligible or even negative financial returns to innovators. Because research funding has significant opportunity costs, investment in these types of commercially friendly strategies will result in fewer resources for social research and public health policies that, although likely to be effective, are unlikely to provide direct economic benefits to those investing in them.

Similarly, in the agricultural sciences, privately funded agricultural research has focused on the development of technologies, such as GM seeds, despite the fact that these technologies are more expensive for farmers and may be less effective at reducing hunger than examining mechanisms that would facilitate food distribution.[37] Such seeds, however, can receive patents, unlike other sorts of interventions that might better address hunger. The same issue arises in privately funded energy research, which has focused on ways to increase energy production from fossil fuels, nuclear, and renewable energy sources, rather than ways to decrease energy consumption or make products more energy efficient, even when these latter strategies could better serve the public good.[38]

Third, the emphasis on commercializing science is more consistent with the funding of research that targets populations with the most resources. In the biomedical sciences, for instance, this translates into the development of drugs and similar interventions directed at those who can pay for them—whether or not the health concerns are important—rather than funding research for conditions that affect poor and marginalized populations.[39] The increasing presence of "me too" drugs—interventions that often do little to improve the standard of care or improve health outcomes—are a case in point. Similarly, lifestyle drugs such as those for baldness or impotence might be effective and safe, but are arguably not responsive to pressing health needs of the global poor, whose health and well-being are likely to be better served by infrastructure interventions that impact the social determinants of health, such as clean water, sanitation, and access to nutritious food, or by low-cost and simple measures aimed at combating diseases that affect them, such as malaria netting.

Insofar as the commercialization of science threatens to undermine the positive role that science can play in improving the lives of all—rather than just a few—it can arguably affect people's perception of the trustworthiness of scientists. This is so, because presumably people trust scientists to produce knowledge that aims to serve the needs of the public rather than simply the good of some. Indeed, according to some evidence, the adverse effects on public trust are already noticeable. Thus, although people in general believe science is a positive force in society, trust in the scientific enterprise is declining.[40] For example, in the United States the percentage of people who have positive views of the role of science in society has gone from 83% in 2009 to 73% in 2014.[41] Similarly, the percentage of those who believe that science has made life more difficult has gone up from 10% to 15% in the same period.

Further evidence that commercialization can have negative effects on people's trust in science comes from surveys and studies regarding opinions about industry. For example, some studies suggest that public trust in university scientists is higher than that of private scientists because scientists publicly funded are perceived to be motivated more by contributing to the public good rather than to profit, and are considered more likely to produce benefits that will be accessible to the public.[42] When specifically asked about the pharmaceutical industry, the majority of European and U.S. patients polled expressed distrust for these companies for a variety of reasons related to the role of science in contributing to the common good, including their failure to assist patients in securing medications, offering drugs with only short-term health benefit, and not serving the needs of neglected patient groups.[43] Given these beliefs, it seems reasonable to assume that the increasing entanglement between industry and academia can cast doubt on scientists' trustworthiness and thus undermine public's trust in science.

Although not without problems, various proposals have been offered to mitigate the negative effects that commercialization of science can have on the direction of research. To the extent that they are successful, implementation of these proposals can have the welcome effect of appropriately promoting people's trust in the scientific community. Most of these proposals have been directed primarily toward addressing the negative impacts of the commercialization of science on health-related issues, but presumably similar mechanisms could be developed to encourage research directed at addressing not only health concerns but also other problems affecting societies, from pollution reduction to the development of environmentally sound housing, to better ways of distributing resources. For example, Thomas Pogge, Aidan Hollis, and colleagues have proposed the creation of a Health Impact Fund (HIF) as a way to address the lack of attention to diseases that affect the poorest countries.[44] The HIF would reward pharmaceutical companies in proportion to the global health impact of their innovations. It would be financed by public funds, with affluent countries contributing more to the pool. Pharmaceutical companies would be invited to register new products, which would entitle them to receive a share of fixed remuneration from the common fund for a defined period. By connecting financial incentives for industry with the global disease burden, the HIF encourages attention to conditions that cause most suffering, rather than to problems that result simply in more profit.

Other scholars have proposed a restructuring of the current funding system in ways that would make scientific innovation more responsive to the needs of all. Some have argued that much scientific research should be socialized and intellectual property rights should be eliminated.[45] In a system of socialized research and no intellectual property rights, financial incentives to focus only on the development of patentable strategies would not exist, and thus scientists would be more likely to seek other types of solutions to the problems that affect societies.

Other solutions involve a restructuring of the current system, albeit a less radical one than socializing research. Some have recommended a gradual reduction of patent life and limits on the kinds of things that people can patent.[46] These reforms could then be combined with a publicly financed global agency that would fund and coordinate research on problems that primarily affect the global poor and promote investigations that attend to strategies other than the development of patentable interventions.

Another alternative would be to fund not-for-profit nongovernmental organizations that would aim at addressing the common good, such as promoting public health.[47] These organizations, some of which already exist and are successful, could be funded through grants, in-kind contributions, and cash donations coming from various sources including public institutions, governments, foundations, companies, and individuals. Rather than conducting the research themselves, these organizations would function as resource allocators, distributing the funds to industry and public institutions for particular types of neglected problems.

As indicated, these strategies have problems and their implementation can be difficult, so we do not mean to suggest that they are either the only options or the best ones. On the contrary, given that all of these strategies have some problems, more research needs to be conducted to explore better measures that can mitigate the effect of industry funding in research agendas.

CONFLICTS OF INTEREST

A second aspect related to the increased commercialization of science that can call into question the trustworthiness of scientist and erode trust in the scientific community relates to the existence of financial conflicts of interest. Disagreements exist about whether Bayh-Dole and similar policies in other countries have truly promoted new discoveries, contributed to bringing more biomedical products to the market, or increased university profits.[48] It is clear, however, that the academic commercial activities encouraged today by most countries create conflicts of interest for both individual researchers and institutions, and thus have the capacity to cast doubt on the trustworthiness of scientists and with it affect people's assessments of whether they should believe particular scientific claims.

In the context of science, a conflict of interest is a situation in which a financial or some other personal consideration (the secondary interest) has the potential to bias, or appear to bias, professional judgment and hinder the integrity of the research or the institutions' responsibilities (the primary interest).[49] These conflicts, both individual and institutional, are created in a variety of ways, including the direct funding of research by industry, consulting fees, serving in advisory boards, holding equity ownership and intellectual property rights, and being a paid speaker for particular companies. The widespread existence of conflicts of interest can actually undermine trust in scientists' claims even when there is no evidence that the conflicts have, in fact, biased research. Their presence can simply create an appearance of bias, and thus can be trust-undermining. An institutional context where conflicts of interest are the norm can undercut scientists' ability to be honest.

Calculating the extent of financial conflict of interests is difficult, but studies have found that such conflicts are common and becoming more so. For instance, a 2003 systematic review assessing the prevalence of conflicts of interest concluded that a 23% to 28% of academic investigators in biomedical research received research funding from industry and about 33% had personal financial ties with industry sponsors.[50] In 2007, another study found that nearly 60% of department heads had personal relationships with industry, serving as consultants, members of a scientific advisory board, paid speakers, or members of a board of directors.[51] It also found that two-thirds (67%) of departments as administrative units had relationships with industry. Similarly, a recent study assessing the prevalence of financial conflicts of interest among panel members producing clinical practice guidelines in Canada and United States found that 48% of authors reported conflicts of interest.[52] Moreover, in spite of the fact that disclosure policies in journals and conference presentations have been standard for some time, evidence also shows that underreporting of existing conflicts of interest is common.[53]

Admittedly, the impacts of conflict of interest on science are also difficult to measure. Nonetheless, various studies have found that conflicts of interest can affect the integrity of the research in various ways.[54] For instance, a variety of studies have found that research receiving funding from commercial parties is significantly more likely to report positive outcomes than studies funded by not-for-profit organizations.[55] Determining the reasons for these results is complex, but some evidence indicates that likelihood of positive outcomes in industry-sponsored studies may result at least in part from methodological choices regarding comparators, dosing and timing of comparisons, selective analysis, and selective reporting.[56] For instance, some research shows that studies sponsored by industry were more likely to use comparators that would produce favorable results, such as examining single interventions owned by the sponsoring company rather than head-to-head comparisons.[57] Likewise, evidence indicates that selective reporting of outcomes is affected by conflict of interest.[58] Conflicts of interest can also negatively affect the quality and reliability of systematic reviews.[59] as well as recommendations in clinical practice guidelines.[60]

Although information about laypeoples' attitudes toward conflicts of interest in science and about the impact on public trust is limited, some evidence indicates that the majority of people agree that scientists should be required to declare conflicts of interest.[61] Also, as indicated earlier, people are more suspicious of the influence of vested interests when research is funded by industry[62] Some data about peoples' attitudes toward conflicts of interest in medicine and biomedical research indicates that people believe that the existence of conflicts of interest decreases the quality of the research evidence.[63]

Insofar as conflicts of interest negatively impact the integrity of the research enterprise, it seems unsurprising that it could affect people's perception of the trustworthiness of scientists. After all, bias in the conduct and reporting of research can have devastating effects on people's health and quality of life. Witness, for example, the efforts by the tobacco industry to fund research undermining evidence of the harms of smoking and second-hand smocking,[64] or those by Merck to understate cardiovascular risks resulting from their drug Vioxx[65] and their adverse consequences of these cases for public health. Clearly, if conflicts of interest can bias scientific results, one is justified in being more vigilant about scientific claims that are produced under conflicts of interest. Indeed, the increased presence of conflicts of interest in science calls into question the trustworthiness of scientists by casting doubts on their honesty and integrity. Similarly, an institutional climate where conflicts of interest are rampant is unlikely to strengthen scientists' resolve for honesty.

The widespread existence of conflicts of interest can undermine trust in scientists' claims even when there is no apparent evidence that the conflicts have in fact biased research. For instance, in 2010, it became public that Rajendra Pachauri, who was then the chair of the Intergovernmental Panel on Climate Change (IPCC), also directed a for-profit renewable-energy company. His company thus stood to benefit from IPCC policy recommendations.[66] Moreover, Pachauri was paid more than a quarter of a million dollars in consulting and speaking fees from other companies that also had a direct stake in the outcome of IPCC climate change policy negotiations.[67] There is no evidence that, in this case, Pachauri's conflicts of interest influenced the content of IPCC reports in ways that would have promoted his own financial interests. Indeed, some studies have shown that IPCC reports tend to adopt overly careful language that underestimate the extent and impacts of climate change.[68] Yet this occurred at a time when the IPCC had no conflict-of-interest policies and their slow response to adopt any such policies in response to the Pachauri case potentially exacerbated distrust of the IPCC, raising concerns both that conflicts could be prevalent among IPCC members and that they were not being managed appropriately.

Concerns about the negative effects of conflicts of interest on the integrity of research has led academic institutions, federal agencies, and publishers to develop policies to attempt to minimize such effects at the individual level. Indeed, federal regulations require that investigators receiving federal grants or contracts disclose significant financial interests to the institution and to the agency. Conflicts-of-interest policies usually include three distinct but related elements: disclosure of conflicts; management of those that are thought to be significant; and prohibition of research activities when such is thought to be necessary to protect the public interest or the interest of the university.[69]

Disclosure policies require researchers to provide information about funding sources, honoraria, consultancies, stock ownership, patents applications, and royalties. Federal regulations impose a disclosure threshold of $5,000 for remuneration and for equity in public companies and no threshold for equity in private companies. Institutions can have polices that are more restrictive. But even if disclosure of conflicts is necessary to attempt to address the concerns that such conflicts raise, it is hardly sufficient.[70] Hence, most research institutions have management plans to deal with conflicts of interest that may unduly influence researchers' judgments. While such plans vary across institutions, normally they involve an oversight conflicts committee that evaluates the nature of the conflict and recommend a particular action plan, such as requiring an independent investigator or committee to evaluate the research related to the conflict, asking the investigator with the conflict to reduce the financial gain, or preventing the researcher in question from participating in particular aspects of the research.[71] The goal of management strategies is to ensure that the risk of bias, or appearance of bias, is limited or eliminated. Management strategies attempt to reduce or eliminate the risk of bias by safeguarding aspects of research that are particularly vulnerable to the (unconscious) influence of financial interests. For example, researchers with significant financial interests in the success of a drug might be prohibited from participating in recruitment for clinical trials so that they do not intentionally or unintentionally select subjects who might be likely to respond to the drugs under study in particular ways. Similarly, conflict-of-interest committees might require an independent investigator or committee to review research design, methodology, or statistical analyses, so as to ensure that research is free from bias. Finally, institutions have policies that prohibit outright particular kinds of financial relationships because they are thought to be excessively fraught with risks of bias.[72] Some of these prohibitions involve a "rebuttable presumption" that allows for exceptions. In these cases, there is a presumption that investigators will be prohibited from conducting research involving human participants when they have a financial stake in its outcome, unless they can show that there is some compelling reason why an exception must be made.[73] The prohibition thus is not absolute and can be rebutted if compelling circumstances make the involvement of the researcher in question justified.

Although development of polices dealing with individual conflicts of interest is now standard, polices that address institutional conflicts of interest are only recently receiving more attention.[74] This is in part the result of the fact that federal granting agencies and journals do not usually have regulations or guidelines for dealing with institutional conflicts in research. Nonetheless, institutional conflicts of interest are not any less concerning than individual ones regarding the integrity of the research enterprise, and thus not any less problematic when trying to promote warranted trust in scientific institutions. Indeed, because institutional conflicts can compromise review and oversight of research, and thus affect many more people, they are even more problematic. Institutional conflicts can involve a disparate and diverse group of individuals who have decision-making authority, from department chairs to deans, vice presidents, presidents, and members of the board of trustees.[75] In spite of this, evidence suggests that only a minority of academic research institutions actually have policies attempting to manage institutional conflicts of interest.[76]

It is not clear, however, that conflict of interest policies are sufficiently effective at the level of either institutions or individuals,[77] and thus may fail to counteract the effects of conflicts on people's suspicions. It is not that existing strategies are useless but, rather, that other mechanisms are also needed. Of course, one obvious solution to the conflicts-of-interest problem would be to prohibit scientists and academic institutions from having any financial stakes in the research they do.[78] Similarly, journals could refuse to publish studies for which any financial conflicts of interest exist. Indeed, many journals have implemented this strategy regarding research funded by tobacco companies.[79]

Although these measures might indeed eliminate the existence of financial conflicts, such proposals are unlikely to be adopted given the decreasing federal resources for research and the increasing emphasis on academic-industry partnerships. Moreover, if as indicated, financial conflicts of interest do not necessarily lead to bias, then prohibiting financial relationships between scientists and private companies altogether might negatively affect scientific and technological innovation, which could also be harmful to the public, since these relationships have contributed to improving public health.[80]

Another possible strategy directed toward limiting the negative effects that conflicts of interest can have on the integrity of the research—and as a result also on warranted public trust—would be to adopt an adversarial system.[81] If such a system were adopted for drug approvals, for example, two groups of advocates, one consisting of industry or industry-sponsored scientists and the other consisting of scientists with no industry funding, would argue on behalf of their respective constituents and respond to various questions, such as whether a drug in question should be allowed on the market, or whether a drug that is already on the market should be allowed to remain so or should be taken off the market. A panel of independent scientists would hear the arguments and make decisions.

Others have proposed the creation of an independent institute committed to neutral hypothesis testing, which would be overseen by a board with members representing different stakeholders.[82] The funding for such institute could be fully public or it could be financed with membership fees coming from the companies that seek to test new drugs. Such a membership fee would be independent of the number and outcomes of the trials conducted by the institute. Either university researchers or the institute scientists could then conduct the research.

Also helpful in trying to reduce the negative effects of conflicts of interest on the integrity of the research and also on trust are requirements to register clinical trials so as to ensure that unfavorable trial results are made public, taken into account, and scrutinized, and drug-related risks are clearly reported.[83] Additionally, increasing avenues for the critical evaluation of research may enable the relevant community of experts to provide a system of checks and balances in helping to prevent the effects of unconscious bias. Greater transparency of data and methodological choices would open up more avenues for critical evaluation by reviewers and other peer researchers. For example, some journals encourage researchers to agree to report more detailed descriptions of trial design, methodological choices, and data interpretation in online versions of their articles.[84] New efforts to create resources to develop and maintain up-to-date information, tools, and other materials to help improve the quality of reporting health research can also contribute to ensure that complete and clear research information is available for scrutiny and that consistent guidelines are used in reporting of health-related research.[85] Recent attempts to address irreproducibility concerns could also prove a reliable tool for verifying the quality and validity of original research.[86]

All these institutional strategies can help support scientists' abilities to be honest. Nonetheless, to be effective in strengthening warranted public trust, the public would need to be aware that these strategies exist and how they function. Awareness that scientific practices are in place to protect the integrity of scientific inquiry can help sustain scientists' credibility.

Scientific Misconduct

In 2011, a Dutch committee investigating Diederik Stapel, a highly regarded social psychologist, on charges of misconduct concluded that he made up or manipulated data in dozens of papers over nearly a decade.[87] The report indicated that many of Stapel's experiments going back to 2004 never took place, and that he gave collaborators made-up data sets. The experiments, which covered topics such as stereotyping, discrimination, and the influence of positions of power in moral thinking, received significant publicity. The committee also found evidence that when Stapel conducted actual experiments, he manipulated the results. According to the committee, many of Stapel's data sets had improbable effect sizes and other statistical irregularities. Nonetheless, his fraud went undetected for years.

In another notorious case of misconduct, Atil Potti, a genomic and precision medicine cancer researcher at Duke University, provided false research data in manuscripts submitted to medical journals and in federal grant applications. An investigation conducted by the Office of Research Integrity concluded in 2015 that Potti had altered data sets, made up the number of patients in various studies, and fabricated their response to treatment.[88] His research used gene expression signatures to purportedly predict a cancer's response to chemotherapy, as well as its risk of relapse. He went as far as to promise subjects enrolled in his chemotherapy clinical trials that he would be able to find the right drug for them with his methods. Duke University recently settled lawsuits brought forward by some of the subjects and their families, who argued they had been entered into Potti's clinical trials fraudulently.[89]

Unfortunately these cases of scientific misconduct are not isolated ones.[90] While determining the prevalence of scientific misconduct is not an easy task,[91] evidence indicates that misconduct is far from uncommon and might be increasing. For instance, a recent study assessing the main reasons for the retraction of scientific publications in the biomedical and life sciences concluded that over 67% of such retractions were attributable to misconduct, including fraud or suspected fraud and plagiarism rather than honest error.[92] The study shows that the incidence of misconduct-related retractions has increased tenfold since 1975. Another study found that retractions resulting from misconduct affect the full spectrum of scholarly disciplines.[93] Moreover, it appears that lower barriers to the publication of flawed articles explain at least in part the increase in the number and proportion of retractions.[94]

Recent meta-analyses of surveys asking respondents about their own and their colleagues' involvement in misconduct have attempted to estimate the prevalence of research misconduct more directly.[95] These studies show that nearly 2% of scientists responding admitted to having fabricated, falsified, or illegitimately modified data, and 28% admitted to knowing colleagues who had done so.[96] Similarly, 1.7% of respondents admitted to having committed plagiarism and 30% said they had witnessed colleagues doing it.[97]

The impact of scientific misconduct is substantial. It can adversely affect the reputation of whole fields of study,[98] waste scarce resources, misdirect research agendas, promote inadequate policies, and result in financial lost.[99] When the misconduct involves clinical research, it can also harm participants in trials and patients treated on the basis of erroneous trial results.[100] Undoubtedly, scientific misconduct can also undermine warranted trust in the scientific community and in the validity of scientific results.

One might object that the existence of scientific misconduct need not damage trust in scientists and in science. After all, the evidence shows that though not uncommon, cases of scientific misconduct constitute a very small fraction of all the research conducted.[101] This should thus increase public confidence that the overwhelming majority of scientists are honest and conduct research responsibly. Furthermore, the growth in retractions of articles appears to be the result of an increased awareness on the side of researchers and journal editors who are getting better at identifying and removing fraudulent papers and those with false results.[102] Insofar as this is the case, then, the public should gain confidence that the self-correcting nature of science is working appropriately. Indeed, at least some evidence suggest that media exposure to cases of scientific misconduct does not adversely affect public trust and might even have a positive effect.[103]

It is true that determinations of misconduct need not erode trust and might even strengthen it. Retraction of publications and findings of misconduct arguably manifest appropriate responses from journals and research institutions, call attention to problems in the scientific record, and promote transparency. Nonetheless, various considerations regarding the existence of scientific misconduct call for caution regarding these considerations and their effect on warranted trust. First, it is not clear that misconduct cases are rare events. As indicated earlier, estimations of the prevalence of scientific misconduct are difficult. There are, for instance, definitional problems related to what counts as misconduct.[104] Thus, some practices such as fabrication, falsification, and plagiarism are widely recognized as clear cases of misconduct and any study assessing misconduct will attend to them. "Fabrication" involves the invention of data sets or results; "falsification" refers to manipulation of equipment, processes, or data such that the research is not accurately represented; while "plagiarism" involves the appropriation of another person's ideas, results, or words without giving appropriate credit. Reports of a low prevalence of scientific misconduct usually involve only these three behaviors. But other forms of questionable research practices, such as improper experimental design and analysis, inappropriate author credit, failure to publish results, refusal to share data, disregard for regulations for the protection of subjects, or failure to disclose conflicts of interest, also exist and are not always recognized as misconduct,[105] even though these practices can be equally detrimental to the integrity of scientific inquiry. Evidence suggests that these other forms of questionable research practices are significantly more common. For instance, asked about whether they or their colleagues had committed research misconduct, over 33% of them admitted to having participated in several of these questionable research practices and up to 72% of them declared they had observed colleagues engaging in questionable research practices other than fabrication, falsification, and plagiarism.[106] Others have estimated that 68% of researchers have inappropriately co-authored papers.[107] Furthermore, many of the studies assessing scientific misconduct prevalence are based on self-reported behavior. Obtaining accurate responses to questions about one's misbehavior is not easy, given that people are reasonably concerned about disclosing unethical, illegal, or otherwise socially unacceptable actions. Thus, it is reasonable to assume that the prevalence of scientific misconduct is likely to be higher than reported.[108]

Moreover, research institutions have a disincentive to establish effective procedures for the detection and documentation of misconduct because their reputation and funding can be severely affected. Indeed, evidence shows that almost 90% of biomedical research misconduct allegations in the United States are dismissed without receiving an initial inquiry or generating a specific record or detailed report to the Office of Research Integrity.[109] Many countries, including European ones, lack national frameworks to deal with allegations of misconduct,[110] and others lack statutory agencies empowered to investigate reports of scientific misconduct and recommend punishment for those found guilty.[111] Furthermore, the majority of scientific journals lack formal misconduct policies.[112] This not only makes it difficult to assess the true prevalence of scientific misconduct but also casts doubts on the ability of relevant institutions to identify, and thus correct, cases of misconduct.

Another challenge in estimating the prevalence of misconduct is the fact that studies focus on academic researchers. But some of the most egregious cases of scientific misconduct are been committed by industry. The tobacco, oil, food, and pharmaceutical industries all have had to pay substantial fines for engaging in practices that involved research misconduct ranging from questionable authorship practices to intentional withholding of safety data, to minimizing risks of consuming their products, to misrepresenting data in clinical and preclinical studies.[113] These cases of scientific misconduct receive a significant amount of media attention, involve systematic and persistent efforts, and have devastating effects on public health.

Second, although identification of scientific misconduct is obviously a good thing, better still, and more likely to sustain public trust, would be the ability to prevent those cases from happening.[114] Some evidence suggests that strong regulatory frameworks, educational measures, and robust mentoring programs can help prevent misconduct.[115] Unfortunately, as we mentioned earlier, strong regulatory frameworks for research integrity are more the exception than the rule, and the implementation of educational and mentoring strategies is spotty in many areas of the world. Additionally, even when policies are in place, research institutions often fail to address questionable research practices and ethical lapses involving their researchers.[116] And evidence suggests that the majority of researchers who observe possible research misconduct are unlikely to report it to their institutional official, for fear that it would adversely affect their reputations.[117] Furthermore, sanctions against those found to have engaged in scientific misconduct are, for the most part, relatively slight.[118] They involve submitting to research supervision and being barred from applying for public funding for a number of years. In particularly egregious cases, it might also involve termination of contracts and academic shaming. Very rarely, misconduct cases can lead to substantial fines to institutions where the fraud is committed and to the researchers. Criminal charges are rare. For example, in July of 2015, Dong-Pyou Han, a former biomedical scientist at Iowa State University in Ames, was sentenced to prison for fabricating and falsifying data in HIV vaccine trials.[119] The case, however, received a significant amount media attention and it drew the interest of Senator Charles Grassley. When pharmaceutical companies manipulate data, fines and marketing restrictions are the usual form of punishment.[120] But even when the fines are substantial, they appear not to serve as deterrent.[121]

Even more worrisome is evidence that factors intrinsic to the current practice of science can encourage outright misconduct and, even more commonly, discourage good scientific behavior.[122] For instance, the pressure to produce has increased considerably over the past few decades. Scientists' careers often depend on their ability to receive grants, publish papers, and more recently, produce intellectual property. At least some evidence suggests that pressure to publish scientific articles contributes to scientific misconduct.[123] These conditions also result in overextended researchers who have less time for proper oversight of research and might become less concerned with responsible conduct of research practices.[124]

Similarly, a research culture that promotes competitiveness—for research grants, prizes and awards, publication in high-ranking journals—might encourage production and creativity, but also likely create an environment where cutting corners pays off.[125] Evidence suggests that competition can undermine responsible research conduct because it contributes to strategic game playing in science, limit free and open exchanges of information and methods, and encourage careless, questionable, and outright unethical research practices.[126]

It might thus well be the case that findings and publicity about scientific misconduct can have a positive effect on engendering public trust in the self-correcting nature of science. Presumably, however, the existence of institutional and structural factors that are likely to encourage questionable research practices, and that make prevention and detection difficult, can weaken scientists' resolve to be honest and can make the development of character traits implicated in trustworthiness, such as honesty and integrity and benevolence, more arduous. Institutional and scientific practices can thus damage the trustworthiness of scientists. Public awareness of the ineffectiveness of institutional practices to uncover, manage, and prevent scientific misconduct can also adversely affect scientists' credibility. Warranted trust in science is thus damaged.

The development and enforcement of research integrity policies would arguably go some way in trying to promote responsible conduct of research.[127] Clarification of misconduct definitions, harmonization of polices, and development of procedures for adjudicating conflicts when harmonization does not occur could limit misconduct cases in international collaborations.[128]

Improving the mentoring system could also be helpful in preventing or limiting misconduct cases.[129] Evidence indicates that trainees who were found guilty of misconduct revealed that their mentors had not established adequate procedures, such as providing clear rules on data ownership and recording and safety, and had failed to review trainees' raw data.[130] Institutions should thus develop and implement policies to ensure that mentors actually advise, teach, and act as role models for their trainees. They should also ensure that mentors are not overburdened with trainees they need to oversee, and they should ensure that mentors with outstanding performances are appropriately recognized and rewarded.

A consistent and strict system of penalties for misconduct, both for individuals and for institutions, and better strategies for detection could also reduce questionable research practices by acting as a deterrent.[131] As we have seen, current sanctions for those found to have conducted misconduct are relatively light, and thus researchers who feel pressure to obtain grants or publish might feel that the risk is low. Stronger whistle-blower policies could also help detection and thus make the risks less worth taking.

Implementation of these and similar strategies is not only likely to reduce the existence of scientific misconduct. By altering contextual factors that weaken scientists' character, enactment of these institutional practices arguably can also create research environments more conducive to the development of trustworthiness. Moreover, ensuring effective communication with the public about the existence and the purpose of these policies can contribute to further scientists' credibility.

Conclusion

There is little doubt that NID can have adverse epistemic and social consequences. Yet the impact that NID can have is likely to be less damaging in a context where people have good reasons to trust scientific communities and few reasons to distrust them. Here, we have called attention to several factors that make NID more likely to grow and take hold, and are more likely to undercut epistemic claims and be corrosive of needed policies and actions. The increased commercialization of science undercuts one of its primary goals—that of contributing to the common good. In a context where economic incentives and profits play a substantial role in determining the direction of research agendas, the common good will take a back seat. But in neglecting science's goal of promoting human well-being, scientific communities give the public grounds to question scientists' benevolence and thus their trustworthiness. Moreover, the commercialization of science results in financial conflicts of interest for both scientists and research institutions, and such conflicts threaten the integrity of research. Although conflicts of interest need not result in biased science, they can certainly lead to it. Because the public lacks the knowledge to determine when conflicts have and have not adversely affected particular scientific claims, rampant conflicts of interest simply create reasonable suspicion regarding scientists' trustworthiness and erode public trust in scientific claims. That scientific misconduct, particularly when understood as including questionable research practices, is far from uncommon can also undermine public trust. This is particularly so when policies to prevent, detect, investigate, and sanction such misconduct are either inexistent or inadequate. Additionally, the current research system is likely contributing to the presence of misconduct cases. All these factors call into question the capacity of research institutions to develop and nurture the trustworthiness of scientists. The present context in which scientific knowledge is produced is, thus, one that undermines trust. Consequently, NID finds a nourishing environment.

Addressing the concerns raised in this chapter will not be simple. The proposals mentioned that aim to eliminate or mitigate the problems discussed here are not without difficulties. Indeed, as we have seen, many of them involve a significant restructuring of scientific practices and institutional policies. Research aimed at better understanding how contextual factors influence scientists' behavior and decision making can provide grounds for developing sound strategies directed at preventing misconduct, limiting conflicts of interests, and minimizing the potential negative effects that current funding mechanism can have on research integrity and direction.

But the need to address these problems is urgent. The distortions resulting from the current research system are not only ethically problematic but also quite likely unsustainable. As challenging as development and implementation of the various strategies discussed here may be, failing to confront these problematic aspects of the research enterprise will lead to a further erosion of public trust in scientific communities and hinder the ability of science to meet its practical aims. Of course, mechanisms that aim to address current problems, even if successful, can only promote trust insofar as the public is aware they exist. Thus, public communication of science not only needs to address scientific content but also ensure that the public is informed about the presence of institutional practices encouraging and cultivating scientists' trustworthiness. More research should thus also be conducted about how best to communicate with the public in successful ways.

The arguments presented here in no way deny the problematic effects that NID can have. On the contrary, it acknowledges them. But insofar as elimination of such dissent is improbable, a focus on the factors that undermine trust in the scientific community and that make NID more damaging can provide us with necessary and important resources in trying to prevent such adverse effects. Tackling these factors can promote the public's trust in the scientific community and the science it produces. It has the added benefit of attending to problems that deserve solutions in their own right. This is surely a complex task, but one that is urgent if we want to ensure that science can fulfill its epistemic and practical goals.

Values in Science and the Erosion of Trust

Recall the case discussed in chapter 6, where emails of climate scientists working at the East Anglia Climate Research Unit (CRU) were hacked and subsequently made public.[1] In one of those emails, widely publicized by the media, climate scientist Phil Jones refers to using a "trick" to "hide" decline in temperatures. He was referring to a statistical method used to exclude certain tree-ring data as proxy data for temperature because, starting around 1960, tree growth slowed in high northern latitudes, making some tree-ring data unreliable as an accurate indicator for temperature after that year. Climate skeptics presented the content of this and other emails out of context, however, and argued that climate scientists were cherry-picking and fabricating data in order to support their desired conclusions.[2] The committee charged with conducting an independent review of the emails cleared the CRU scientists of any scientific misconduct, any misleading use or exclusion of tree-ring data, or any other behavior that would have undermined the conclusions of the IPCC's assessment reports or the advice given to policymakers. However, they also found "a consistent pattern of failing to display the proper degree of openness," on the part of both CRU scientists and the University of East Anglia (UEA).[3] The committee also concluded that CRU scientists and the UEA were generally unhelpful in responding to information requests related to the investigation and that there was "evidence that emails might have been deleted in order to make them unavailable should a subsequent request be made for them."[4] Thus, even if the science itself was sound, this episode cast doubts on the trustworthiness of climate scientists.[5] Repeated failures to be transparent and open contributed to a public perception of climate scientists as dishonest, and generated concerns that they were motivated by a political agenda that led them to fudge the data and suppress dissenting views in ways that would compromise the epistemic integrity of their research.

While the so-called climategate case was one of the most publicized, it is not the only instance where scientists have been accused of allowing their political values to illegitimately influence their research. In 2012, the journal

Food and Chemical Toxicology (FCT) published a study by French molecular biologist Gilles-Éric Séralini that reported higher rates of tumors and kidney and liver damage among rats fed Bt maize and the herbicide RoundUp than rats in control groups.[6] The two-year-long study reported the most striking results among female rats, who died at a rate two to three times higher than the respective control group animals. Séralini was also the president of the scientific advisory board for the Committee of Research and Independent Information on Genetic Engineering (CRIIGEN), a group that opposes genetically modified foods. He is also the author of a book and a documentary critical of this technology.[7] Shortly before publication, Séralini held a press conference to publicize his findings, but he required journalists to sign a confidentiality agreement that imposed strict constraints on disclosing information about the study.[8] As a result, the first round of stories on Séralini's findings were not accompanied by any critical feedback from other experts in the field.[9]

Once the findings were published, numerous scientists and scientific groups criticized the study thoroughly and argued that it was methodologically flawed. They raised objections about the use of a rat model that was known to be highly susceptible to tumor development and other health problems, the small sample size, the statistical analysis, and the lack of information on key experimental features, such as how much food and water the rats were given.[10] Importantly, several argued that Séralini's research had been driven by a political environmentalist agenda that had biased the experimental design and the study results.[11] The journal subsequently retracted Séralini's study without the authors' consent. The decision was based on the fact that *FCT* found the results inconclusive rather than on a finding of misconduct.[12] Shortly after the authors published a slightly revised version in *Environmental Sciences Europe*.[13] They also accused many of their most ardent critics of being associated with Monsanto and having conflicts of interest that they failed to disclose.[14] Concerns about the influence of political agendas on studies regarding the safety of genetically modified products (GMPs) have also contributed to skepticism among the public about the credibility of this area of research.[15]

These and similar cases call attention to another factor that arguably casts doubts on scientists' trustworthiness and undermines public trust in science: the belief that the results of controversial areas of inquiry, such as climate change, vaccine safety, and GMP research, are biased because they are influenced by scientists' values. In this chapter, we explore how beliefs about the role of nonepistemic or contextual values in scientific inquiry can lead people to question the trustworthiness of scientists and undermine trust in scientific claims. As we discussed in chapter 7, normatively inappropriate dissent (NID) is likely to have more damaging epistemic and social effects in a context in which people are suspicious of scientists and their claims. We thus also propose some strategies to address legitimate concerns about the role of contextual values in science that aim to build and sustain warranted trust in scientific communities.

Values and the Epistemic Integrity of Research

The authority that scientific inquiry enjoys in the public sphere stems in great part from the ideal view that science is objective—that is, that its methods and findings are not influenced by particular perspectives, political agendas, value commitments, personal interests, and the like. When the public perceives that scientists have allowed personal, political, ethical, or social values to influence their research, it is likely to believe that the epistemic integrity of the research has been compromised. In other words, one might think that science must be disinterested in order to be objective. However, evidence shows that a significant amount of the public believes that research findings on climate change, for instance, are influenced by scientists' political leanings.[16] This arguably explains, at least in part, why the majority of Americans are skeptical of climate scientists.[17]

But, as discussed in chapter 3, science need not be wholly disinterested in order to be objective. As we have seen, many areas of scientific investigation are motivated by ethical concerns, such as protecting human health and well-being. Insofar as members of the public believe that interested science is necessarily biased, facilitating and sustaining trust would require correcting this false assumption.

Yet people might accept that values play a significant role in what scientific communities study and why they do so, and they can nonetheless be concerned that scientists will allow their moral or political values to influence their reasoning in more substantive ways. They might worry that scientists' personal values can, even unconsciously, skew scientific reasoning in the collection, characterization, or interpretation of evidence in order to arrive at a particular favored conclusion. This would amount to wishful thinking, as explained in chapter 3—that is, to accepting theories because one wishes them to be true, rather than on the basis of evidence that they are true.[18]

This concern is reasonable. Values can sometimes influence science in ways that lead to biased reasoning. The case of Trofim Lysenko is often taken to be a paradigm for how political values can thwart the integrity of science. Lysenko, a Soviet agrobiologist, was investigating ways to increase crop yield at a time when Joseph Stalin was collectivizing Soviet agriculture.[19] Pressured by Stalin to produce rapid results, Lysenko claimed to be able to increase crop yields dramatically by altering environmental factors in seed planting, such as light, temperature, and moisture. Many at the time accepted these claims, even though they challenged accepted principles of genetics, because they cohered with prevailing Marxist ideologies about the importance of environmental factors over inherited traits. Moreover, in rejecting accepted genetic theories, Lysenko was championed as a working-class peasant who was challenging the intellectual elite.[20] Nevertheless, his rejection of genetics was unfounded, and later many came to view his work as "pseudoscience."[21] Lysenko's commitment to promote or retain a Marxist ideology prevented him from producing epistemically sound research.

Similarly, many scholars have documented instances of sexist, androcentric, heterosexist, racist, and classist assumptions that adversely influenced the epistemic integrity of science.[22] For instance, biologists investigating the role of sperm and egg in reproduction relied on gendered stereotypes when characterizing the sperm as actively "burrowing and penetrating" into a passive egg.[23] Their research thus failed to account for chemical properties on the surface of the egg that play an active role in attracting and fusing with the sperm during fertilization. Scientists' values and background assumptions about gender distorted their perceptions of a process that was mutually interactive. Scientists observed instead what they expected to see. Similarly, during the nineteenth century, scientists' assumptions that women and African Americans had lesser intellectual capacities led them to discount data that contradicted their preferred hypotheses regarding the superiority of white males.[24]

In addition, social or political values can lead individual scientists to engage in hype, or present scientific findings as being more justified or wider-reaching than is warranted by the evidence. Climate skeptics, for instance, fear that climate change scientists with an environmentalist agenda overstate the certainty of climate change impacts and incorrectly label them as "catastrophic."[25] Climate scientists, however, defend their conclusions because they believe the stakes are high and the potential outcomes appropriately seen as devastating. Similarly, skeptics have accused climate scientists of communicating results in ways that neglect or downplay uncertainties, while many denunciate skeptics for exaggerating both the degree and the significance of particular uncertainties related to climate change.[26] The question of how boldly or conservatively to present results is not determined by science alone. This, however, gives rise to fears that scientists' political values will motivate them to overstate or understate their results in ways that are inconsistent with existing evidence.

Trusting the Epistemic Integrity of Research

It seems, then, that scientists' political and social values can sometimes lead them, intentionally or not, to produce or report scientific conclusions that are epistemically unsound or unjustified. Clearly, those who present dissenting views on areas such a climate change or vaccine research often use this argument to cast doubts on the credibility of scientific results in these areas.[27] Similarly, proponents of the consensus view on climate change and vaccine research argue that contrarians allow their political ideologies to bias their claims.[28] That political ideology tracks people's concerns about the consequences of climate change, their confidence in climate scientists' competence, and their beliefs about the objectivity of the research constitutes evidence that concerns about the effect of political values in science can undermine people's trust in science.[29]

What should be done about these reasonable concerns? What strategies can be implemented that could assuage people's worries about illegitimate influences of contextual values in science that would contribute to create and sustain warranted trust in scientists? Given concerns about the ways in which values might compromise the integrity of scientific research, it might be tempting to think that a good solution would be for scientists to attempt to more rigorously adhere to a value-free ideal (VFI) of science. Recall from chapter 3 that, according to the VFI, scientists, *qua* scientists, ought never to rely on ethical, political, or personal value judgments when evaluating or justifying scientific theories.[30] While there may be some scientific decisions that involve social, ethical, or political considerations, such as which research programs to pursue or whether some methodological practice conforms to standards for the responsible conduct of research, supporters of the VFI maintain that such values should never play a role in the evaluation or acceptance of hypotheses. Thus, some have argued that scientists must be careful to distinguish their roles as scientists from their roles as policy advisors.[31] Advocating for particular policies may require endorsing values, but this should be distinct from scientists' role in generating evidence or in justifying particular scientific claims. Practicing scientists should thus avoid politicizing science so that they can maintain warranted public trust and protect their role as honest brokers of the facts.[32]

Yet, proposing the VFI as a solution to concerns regarding the potential negative effects of contextual values in science is problematic, for several reasons. First, as many have argued, adhering to this ideal is simply not possible.[33] As noted, the influence that these values have is often unintentional, such that even conscientious scientists who do their very best to adhere to the VFI may not be aware of what values are affecting their reasoning. Moreover, value judgments are inevitable, given the uncertain and pragmatic nature of science.[34] Scientists rarely, if ever, achieve certainty regarding scientific claims, and thus they must make judgments about what outcomes to accept or endorse in order for scientific investigations to proceed. But in making judgments about whether evidence is sufficient to accept or reject a hypothesis, model, or interpretation of data, scientists cannot avoid being affected by their value preferences.

Second, the VFI is inherently flawed because value judgments play important and beneficial roles in scientific reasoning.[35] Some of the ways in which values can play a role in scientific reasoning were highlighted in chapter 4. For example, sometimes the content of scientific theories involves normative or value-laden concepts.[36] Research aiming to measure harms, impacts, or risks, for example, clearly involves normative concepts that rely on assumptions about what we take to be central to well-being or about what human or non-human interests need protection.[37] In measuring climate impacts, whether the loss of a language or cultural tradition counts as an "impact" depends on judgments about what goods are worth protecting. In biomedical research, what constitutes a "side effect" or an "adverse outcome" that ought to be measured or reported in clinical trials relies on value judgments about what conditions we take to threaten human well-being.[38] In conservation biology, employing concepts such as "ecological restoration," "sustainability," "healthy forests," or "ecosystem integrity" presupposes values about what we take to be important to protect or what we believe is central to environmental flourishing. Scientific claims are thus not merely descriptive claims. But if that is correct, then value judgments will be relevant to decisions about how scientists should test such hypotheses and what exactly they should measure. Eliminating such values would thus have a negative impact on the epistemic quality of scientific research.

Even when the concepts in scientific theories appear to be descriptive, the choice of which conceptual frameworks to employ can depend on less obvious contextual value judgments.[39] For instance, how many orchids in Indonesia a scientist considers to have become extinct will depend on criteria for species membership. Because there is not one universally accepted set of criteria, biologists will disagree on the precise number of orchid species. If two biologists use different criteria for species membership of orchids, they can also end up disagreeing on the number of orchid species that have become extinct in some particular area. Yet, which ontological framework to use in order to understand species depends on value judgments about what particular categories will help make visible those features that are of the most interest, depending on the research context and aims—including social aims—of the research.

In addition, contextual values can be relevant to determining the methodologies best suited to acquiring data. Consider research on the toxicity of Bt genetically modified maize. Persistent disagreements exist about what would constitute the right kind of empirical data to justify or challenge claims about nontoxicity. The de facto, established methodological norm in such studies is to use an extraction of the purified Bt protein taken directly from the bacteria rather than using a protein from a genetically modified maize plant.[40] But these different methodological approaches will produce empirical evidence about either the toxicity of the Bt protein by itself or the potential toxic effects caused by the genetic modification of the maize plant.[41] Which methodology is better suited to produce the sort of evidence that could inform sound public policy is thus not given by nature; it requires value judgments about what we want to measure. Similarly, methodological decisions about the duration of studies, the appropriate make-up of test subjects, and the selection of biological endpoints depend on contextual value judgments about the social aims of the research.[42] In such cases, the VFI would not be desirable because it would require scientists to ignore considerations relevant to their decision making. Appealing to contextual values can sometimes promote the epistemic aims of the research.

Value judgments related to endorsing and advancing particular social and policy aims of research can also be relevant to employing and adjudicating between traditional epistemic and cognitive values.[43] Consider, for instance, models in economics that relied on "heads of household" as the only economic actors in families. Such models involved an implicit epistemic tradeoff.[44] They posited fewer causal actors, and thus were useful because of their simplicity. However, they also obscured our understanding of certain kinds of economic transactions and inequalities, particularly gendered inequalities within families. Making such inequalities visible required more complex models. Whether one should prefer simpler or more complex depends on contextual value judgments about how important it can be to make such inequalities visible.

Similarly, climate change models are to some extent "tuned" or adjusted to better match established empirical observations.[45] But, there are different features of the climate system that model tuning might strive to address. Tuning to make a model more consistent with observed data in certain respects often means that it is less accurate in others. For example, models that are tuned to better represent the distribution of tropical precipitation between land and ocean in maritime Southeast Asia may less accurately represent tropical intraseasonal variability.[46] Thus, decisions about how to fine-tune models often involve judgments about which features of the climate system we think are most important to represent correctly. Such decisions, however, depend on value judgments about what is most important (e.g., whether we are more concerned with generating predictions about the longer-term distribution of precipitation or week-to-week extreme weather events such as monsoons). Which tradeoffs are justified depends, in part, on the sorts of predictions or trends that are important to advancing policy aims.[47]

Finally, value judgments play important roles in assessing risks related to collecting and characterizing data.[48] In climate research, cloud formation represents a significant area of uncertainty.[49] Clouds, depending on how they are formed, can either amplify or dampen global warming. Yet, despite the uncertainties, scientists must make some assumptions about how to represent cloud feedback in order to run models.[50] Whatever assumptions they make, they run the risk of being wrong. Scientists must make judgments, then, about which sorts of errors are acceptable. Yet, such judgments depend on how bad scientist perceive the social consequences of error to be. On the one hand, falsely assuming that cloud conditions will amplify warming could lead to overestimating the extent of global warming and thus to overregulation of

CO_2. On the other hand, falsely assuming that clouds will lessen warming could lead to underregulation of CO_2 and may lead to greater climate change impacts than predicted. This could have potentially devastating and irreversible consequences for biodiversity, human health, food production, and economic systems.[51]

As this overview shows, contextual values play necessary and important roles in collecting and interpreting data and in evaluating hypotheses. Promoting adherence to the VFI would hinder the scientific and social aims of research, at least in some cases. A strategy that undermines the epistemic quality of scientific research is clearly inadequate as a way to promote warranted trust in scientific communities.

But if promoting adherence to the VFI is not a solution, what can be done to address the legitimate concern that contextual values *can* lead to bias so as to facilitate warranted public trust? Although it is neither possible nor desirable to prevent such values from operating in science, scientific communities can be organized in ways that minimize potential biases and that increase the objectivity of science. For example, creating public avenues for critical evaluation of research can help identify illegitimate uses of contextual values. When appropriate venues for criticism of scientific reasoning exist, members of the scientific community have the ability to point out problematic background assumptions or methods employed by other scientists. Mechanisms such as peer review are clearly important for protecting the epistemic integrity of research. It might be difficult for individual scientists to recognize when their own reasoning is biased, but other scientists with the appropriate expertise can often do so. Recent proposals to address concerns about reproducibility are also likely to strengthen scientists' ability to identify flawed research.[52] Requirements to provide clear information about crucial experimental design elements such as blinding, randomization, replication, sample-size calculation, effect of sex differences, strains of animals, reagents, or statistical methods can facilitate review and critical evaluation of research prior to its dissemination. Making all relevant research data open once the research is published also facilitates ongoing critical evaluation of scientists' methodological choices and approaches.[53]

Granted, as discussed in chapter 2, these mechanisms will only be effective in identifying bias if the values guiding the research in question are not widely held by scientists. As we have indicated, value judgments often operate as background beliefs in decisions throughout the research process. If those values are common, then their influence will likely go unnoticed. Attending to the potential biasing role of contextual values in science thus requires of scientific communities more than the creation of public venues for criticism. It also necessitates that such communities embrace diversity of values and interests among participants so as to maximize the chance that erroneous value judgments are identified and challenged.[54] When values are different from one's own, it is easier to identify their effect on scientific reasoning.

Diverse scientific communities can be useful not only in minimizing bias in evaluating hypotheses but also in preventing bias regarding the framing of research problems and the direction of research agendas. Insofar as interests shape what scientists study and why they study it, incorporating members who hold diverse interests can broaden the range of phenomena for investigation, as well as increase the creativity of scientific communities.[55]

Strategies to promote pluralism of methodological approaches and ideas can also be helpful in addressing concerns about the negative influence of contextual values on the epistemic integrity of science.[56] Such pluralism can encourage the simultaneous pursuit of different research agendas, making it more likely that scientists will investigate different dimensions of scientific problems. This could counteract the problematic influence that contextual values can sometimes have on the direction of research. Moreover, insofar as scientists use different methodological approaches, this can lessen any inappropriate effects of value judgments in methodological choices.

It is beyond the scope of this work to argue what specific mechanisms will be more effective in enhancing the epistemic integrity of research. But building and sustaining warranted public trust in science requires that we attend to legitimate concerns about the negative role of contextual values in research. Doing so involves organizing scientific communities and developing research practices that protect the epistemic integrity of research. It also requires effective communication with the public both about the necessary and important roles that contextual values play in knowledge production and about the existence of mechanisms aimed at limiting illegitimate influences of political agendas in science. These strategies are beneficial insofar as the promote research integrity and facilitate well-grounded public trust in science. But as we have argued, an environment where justified trust in scientific communities is robust is also one where NID is less likely to produce damaging epistemic and social effects.

Whose Values? Trusting Values in Science

But wouldn't the recognition that contextual values play a role in science, even if a legitimate one, actually undermine appropriate trust in scientific communities rather than facilitate it? After all, value disagreements are common and members of scientific communities, even if diverse, can still hold values that clash with those of other stakeholders. Indeed, evidence shows that, at least for some controversial policy-relevant scientific areas, political ideology is a significant factor in people's degree of confidence about scientific testimony.[57] Thus, even if mechanisms exist that facilitate public trust in the integrity and reliability of scientific research, people might still be concerned that value-driven research is partial, to the *wrong* values insofar as such values are not shared by members of the public evaluating research conclusions. We saw how this occurs in the case of commercial research. Research funded by pharmaceutical companies can produce medical interventions that are safe and effective. Nonetheless, the problems that such interventions try to address might not be the most pressing from a public health point of view, or may be less effective than preventive strategies that would be less profitable.

Other sorts of contextual values can also affect research in a variety of ways, as we have seen. First, they can determine how research problems are framed. Consider, for instance, research on contraceptive methods. Until recently, research aimed at developing contraceptives has largely focused on women.[58] Unsurprisingly, while multiple reversible methods are available for them, from barrier to hormonal ones, only condoms are available for men. Gender biases regarding who is responsible for contraception are arguably at play in the framing of contraceptive research,[59] which means women carry the burden of testing and using contraceptive measures. Even if existing contraceptive methods are safe and effective, still many women—and men—can be reasonably concerned that the values that underlie the framing of this research do not promote their interests.

Second, contextual values play a role in determining what sort of data to collect, and hence what evidence will ultimately be available for justifying particular public policies. While most would recognize that climate change research should provide us with information that will allow the protection of those things we care about, what is of value to climate scientists might be of little interest to other communities or even positively rejected by them. For instance, some have argued that the interests of those in the Global South have been disregarded by current climate change research.[60] There are multiple ways of modeling impacts that are all empirically adequate, but they attend to and neglect certain aspects of the climate system. For example, integrated assessment models commonly measure climate change impacts in ways that focus on certain easily quantifiable goods, such as loss of life or property, while they neglect other things, such as loss of cultures or cultural practices.[61] Thus, people in the Global South might be disinclined to find those models reliable, not because they believe them to be epistemically unsound but because they believe, in some cases correctly, that their interests have not been taken into account.

Third, value judgments underlie risk assessments about whether theories and models are good enough to make accurate predictions for grounding public policy. Consider, for instance, questions about the burying of nuclear waste. In assessing whether or not Yucca Mountain could safely contain nuclear waste, hydrogeologists had to make judgments about whether their data were sufficiently reliable. While scientists were aware of several methodological problems and limitations with their models, and also knew that there were significant health and environmental risks at stake, they agreed that the models were the best available and "good enough" for basing policy decisions.[62] But those living in the Yucca Mountain area could reasonably be skeptical about scientists' willingness or ability to protect their communities' interests when assessing risks.

If contextual values can influence scientific inquiry in ways that prioritize some interests and values over others, or can fail to be fully responsive to the concerns of all stakeholders affected by research, then it is reasonable for people to be skeptical of some scientific testimony. Would this mean that public trust in scientists and their testimony will be dependent on whether people agree with the values that guide the research in question?[63] It is not clear. On the one hand, as we indicated earlier, at least some evidence suggests that agreement or disagreement with the perceived ideological commitments of climate scientists, for instance, tracks agreement or disagreement with the conclusions of such research.[64] On the other hand, trust is a complex phenomenon and, as we have seen, multiple factors appear to play a role in whether people believe the testimony of scientists is reliable.[65] It is also reasonable to think that insofar as people believe that scientists are trustworthy, they can believe their testimony is reliable even if those who have different values ultimately disagree with some of the conclusions. That is, people who perceive their values to be different from those guiding particular research programs can trust that scientists have not inappropriately biased their conclusions and at the same time can disagree with the conclusions in question, given that those conclusions would be different had the research been guided by other contextual values.

More research is thus needed to ascertain how peoples' perceptions of scientists' values influence their trust in the scientific community and how important agreement with scientists' values is for trusting them. But whether or not such agreement is necessary, various strategies can be helpful in promoting warranted public trust in science related to the necessary role that contextual values play in scientific research.

First, even if science is necessarily driven by *some* contextual values, we should not leave decisions about which values should be endorsed to scientists alone. Allowing scientists alone to make ethical, political, or social value judgments when conducting research gives them disproportionate power in shaping the science that will be available to inform policy decisions.[66] Scientists have no special expertise or authority in making these judgments, and thus, it is unclear that they alone should be the ones deciding what values to endorse. Moreover, as a group, scientists are unrepresentative of the diverse stakeholders affected by science. In addition, there can be some reasonable disagreements about social, political, and ethical values, and in pluralistic societies all stakeholders should have equal representation in determining which values to endorse or prioritize in cases of conflict. In democratic societies, legitimized institutions, and not a handful of unelected scientists, should be the ones deciding on collective goals and values.

Second, strategies directed at ensuring a fair representation of the various values or interests that can be at stake when conducting scientific research can also be helpful in facilitating trust. One such strategy would be to increase transparency both about the decisions scientists make that involve value judgments and about the particular values that underlie such judgments.[67] (Making values explicit to the public and policymakers allows stakeholders to assess for themselves whether the value judgments are justified. For example, climate scientists may decide that, in the face of uncertainties, it is better to run the risk of overestimating climate change, so as to protect against the worst-case scenario. Therefore, they may endorse assumptions in climate models that risk overestimating the extent to which average global temperature will increase. Making these assumptions transparent—both the value judgments about which risks are acceptable and the decisions for which there was significant uncertainty—would allow policymakers to evaluate how different value judgments might have produced different scientific results.[68] Such assessment could help policymakers and members of the public to determine whether to accept or reject the scientific conclusions presented. Notice that transparency can facilitate trust in scientific communities without leading to agreement with the particular scientific conclusions because those evaluating the research do not share the values that have guided it.

But transparency about values by itself is insufficient to address concerns about the multiplicity of values or interests that stakeholders can have.[69] On the contrary, transparency would bring to light the fact that scientists' interests can be inconsistent with those of at least some stakeholders. So strategies are also needed to effectively incorporate stakeholders input in determining which values ought to guide scientific inquiry. One such mechanism could be to involve stakeholders in establishing the aims of research.[70] This could happen in the early stages of developing research programs, such as when advancing calls for grant proposals. If research programs have clearly defined social goals that are justified by a representative range of stakeholders, scientific decision making could be justified in relation to those democratically endorsed goals, rather than the goals of particular researchers.

Indeed, there are many examples where stakeholder input has been successfully incorporated in establishing and refining aims of research and involving citizens in research.[71] Community-based participatory research, for instance, has been used in various research areas, where scientists work with community advisory boards (CABs) with representatives of groups affected by the research.[72] This has been common, for example, in both national and global research on HIV/AIDS prevention and treatment.[73] In these cases,

CABs participate not merely in crafting policy recommendations for, for example, needle exchange or HIV education programs. Rather, they play a role at various stages throughout the research process: in formulating what the policy aims and the priorities of the research should be; giving feedback on the extent to which methodological decisions sufficiently advance those aims, such as clinical trial methodology; and providing critical feedback on assumptions that scientists have made in interpreting data. Similarly, in the context of climate change research, there are increasing efforts to incorporate stakeholder input throughout the research process.[74] The UK Climate Impacts Programme, for instance, has developed mechanisms for working with stakeholders to identify adaptation needs and receive critical feedback on modeling strategies to produce more useable knowledge.[75]

Increasing diversity among scientists may be helpful not just in minimizing biases but also in broadening the range of interests and values that can legitimately guide scientific research. Scientific communities should thus encourage the participation of scientists who are socially diverse or represent a diversity of life experiences.[76] The creation of science advisory committees constituted by scientists with varied political values could also contribute to value pluralism.[77] Given that the products of science often have different consequences for those in different social circumstances, increasing such diversity within scientific communities is likelier to ensure a better representation of the interests of all relevant stakeholders.

More work needs to be done to identify what mechanisms might best engage stakeholders in science and be able to attend to various legitimate interests. Our claim, however, is that conducting research aimed at identifying and implementing such mechanisms is crucial in building and sustaining warranted trust in scientific communities and their testimony. Ultimately, such trust could help limit the damages that NID can have.

Conclusion

Concerns about the inappropriate role of contextual values in scientific research can lead people to be suspicious of scientific communities and their testimony, making NID more damaging than it otherwise would be. The presence of social, political, or other values can undermine the epistemic integrity of the research. When the public believes that scientists have a political agenda that influences their research, it calls into question their trustworthiness. It can be tempting, then, to promote adherence to the VFI of science as a way to address this source of public mistrust. Nonetheless, although undeniably contextual values negatively influence science sometimes, they also play necessary and important roles in scientific reasoning. Therefore, requiring scientists to abide by the VFI is misguided. Facilitating warranted public trust calls for strategies that are likely to reduce an illegitimate use of contextual values in science. Institutional practices, such as promoting public avenues for criticism, increasing diversity among scientists, and encouraging pluralistic approaches to research that increase opportunity for identifying and critically evaluating values or correct for any negative or biasing influence such values may have are likely to be more successful both in ensuring the epistemic integrity of the research and in creating and sustaining justified public trust in scientific communities. These sorts of mechanisms, however, are not sufficient to address reasonable concerns about what particular values should be prioritized. Even if contextual values do not undermine, and can actually improve, the epistemic soundness of research, decisions about what values to favor should hardly be left to scientists alone. Scientists have no expertise in this area and their choices are unlikely to be representative of the diverse values that exist among stakeholders. Thus, creating and sustaining warranted public trust requires the implementation of strategies that promote transparency of values and that are inclusive and representative of the diversity of stakeholders' interests. It is true that implementation of these strategies can make it even more difficult to build consensus for particular policies, as it may be challenging to agree on what values to promote or not. But disputes about what policies are best that result from genuine disagreements about values are part and parcel of robust democratic societies.

10

Where Disagreements Can Lie

ATTENDING TO VALUES IN POLICY

Since 2015, individual E.U. countries can decide to ban genetically modified products (GMPs), even if the European Union's own regulators have declared them safe for cultivation. Countries such as Germany, France, Hungary, Greece, Latvia, and Scotland prohibit genetically modified crop cultivation on their country's fields. Critics of the strict regulations present in many European countries argue that such policies illustrate a clear disregard for the scientific evidence and for what critics argue is a broad scientific consensus regarding the safety of GMPs.[1] Prohibitions against these products, they argue, constitute pandering by officials to various pressure groups.[2] For these critics, the evidence for the safety of GMPs necessarily calls for policies that allow for cultivation and commercialization.

Similarly, the U.S. Food and Drug Administration (FDA) recommended in 2011 that the emergency contraceptive drug Plan B be available over the counter to all females of child-bearing age, including those under seventeen years of age, who until then needed a prescription. They based the recommendations on an analysis of scientific evidence that the drug was both safe and effective, and that it had met all the FDA criteria for over-the-counter availability. In an unprecedented move, however, then-secretary of Health and Human Services, Kathleen Sebelius, overruled the FDA's recommendation.[3] She was accused of ignoring the scientific evidence and her decision was decried as another occasion where the "politics of birth control have trumped science and sound public policy."[4] In 2013, a federal district judge in New York, Judge Edward R. Korman, overturned the Obama administration's ban. His decision was hailed as triumph of science over politics.[5] For critics of Sebelius, the science clearly dictated that Plan B be available over the counter without age restrictions.

The Paris Agreement, the first-ever universal, legally binding global climate deal reached in December 2015 by 195 nations, aims to reduce greenhouse gas emissions to keep warming below 2°C over the next century. Accomplishing this would require large emitters such as the United States to reduce greenhouse gas emissions to nearly zero by the year 2060.[6] During the 2016 presidential campaign, then-candidate Donald Trump announced his intention to pull out of the agreement,[7] as well as strike down any regulations related to emissions imposed by the Obama administration, calling climate change a "hoax" perpetuated by the Chinese government. Consistent with these promises, Trump subsequently nominated Scott Pruitt to direct the Environmental Protection Agency (EPA). Pruitt has argued that emissions regulations are unwarranted because "scientists continue to disagree about the degree and extent of global warming and its connection to the actions of mankind,"[8] suggesting that his opposition to climate change policies is related to uncertainty about the science. His critics argue that he is ignoring the scientific facts and the overwhelming scientific consensus about anthropogenic climate change.

What these cases have in common is an implicit assumption about the role of science in public policy. Criticisms that policymakers' actions ignore science presuppose that the reliability of an empirical claim—that GMPs are safe for consumption, that Plan B is safe and effective for younger females, or that anthropogenic climate change is occurring—calls for a particular policy decision: that GM crops should be cultivated and marketed, that Plan B should be available over the counter to all women, and that climate change mitigation policies should be implemented. That is, in many controversies, both sides appear to assume that settled science would be sufficient for determining policy. Consequently, they take their disagreement to be a debate about whether, in fact, the science is settled.

Although the frequency of these sorts of cases indicates that this assumption is common, when explicitly stated, few would be willing to defend the claim that science is all that is needed in public policy decisions where scientific evidence is pertinent. While science is necessary to inform many public policies, science alone does not dictate which policies should be implemented. A variety of value considerations (political, social, and ethical) are necessary and appropriate when making policy decisions for which science is also important.

A reason why this incorrect assumption is so common is that, in many cases, the values that underlie policy choices are shared ones. For instance, if scientists find that a medical drug is ineffective or unsafe, then it seems that the evidence straightforwardly dictates policies that restrict or ban the drug in question. If they determine that a substance is extremely toxic when consumed by people, scientific evidence appears to dictate policies that would limit its use. But the direct path from evidence to policy is an illusion. Relatively uncontroversial value judgments accompany these policy decisions, among them that human health ought to be protected. Because the values are widely shared, they go unnoticed, but empirical facts alone do not dictate policy decisions.

The fact that so many seem to hold the implicit assumption that science alone calls for particular policies is not problematic merely because it is false. It also has negative consequences for policy debates: it diverts attention away from the values that necessarily play a role in policymaking and prevents assessment of such values, some of which might be illegitimate or unjustified.[9] Furthermore, when debates about policy focus solely on what the scientific evidence does or does not show, they neglect what is often the real cause of policy disputes: disagreements about the underlying values at stake.

In previous chapters we have seen that many contend that resistance to particular policies and actions is the result of problematic dissent that confuses the public and policymakers about the state of the scientific evidence and may even erroneously lead them to doubt that a scientific consensus exists. Failures to support emissions regulations, widespread use of GMPs, or mandatory pediatric vaccines can indeed result from such dissenting views, as well as from their dissemination by the media and by powerful private interest groups with significant financial resources.[10] Insofar as resistance to particular policies is the result of confusion about the state of science, we have argued that we ought to focus on ways to facilitate warranted epistemic trust between laypersons and experts. In this chapter, however, we wish to challenge the assumption that policy disputes are largely the result of misunderstandings about particular empirical claims. We argue that resistance to science-based policy recommendations can also arise because of disagreements about the values that underlie policy choices, rather than simply because of confusion about the state of the science.[11] The focus on public misperceptions about the science obscures the fact that people can also fail to support certain policies or act in particular ways not because they erroneously reject the scientific consensus or have false beliefs about the scientific evidence—that is, not because NID produces confusion or misunderstanding. Rather, their rejection of such policies or behaviors can be the result of disagreements about the values that ground the policies or behaviors in question. For instance, social science evidence suggests that people who have a strong sense of social justice or have an egalitarian worldview are more likely to support climate change mitigation policies than those who give more weight to values such as wealth.[12] Similarly, perceived costs or threats to other values has also been a factor in public rejection of particular climate policies.[13] Whether such rejection is justified or not will thus depend on the reasonableness of the values in question rather than on disagreements regarding the scientific evidence. To the extent that opposition to adopt or accept certain policies is the result of disagreements about values, the exclusive attention to doubt and confusion about science as the culprit of stalled policies is misguided.[14] If so, understanding the ways in which divergences in values can influence policy debates is crucial to being able to advance such debates in fruitful ways. Thus, in this chapter we offer a second constructive recommendation for moving laypersons to support science-related policies: an increased focus on potential differences in the values that underlie policy decisions.

In what follows, we identify different ways in which value disagreements, rather than scientific disputes or confusion about empirical evidence,[15] might reasonably lead people to disagree about whether to reject or support particular policies or actions. Specifically, we argue that failure to accept certain policies or actions may be the result of disagreements about what has value, how to interpret particular values that are shared, how to weigh competing goods when they conflict, or how best to promote particular policy goals. These value disagreements are not mutually exclusive and, as we shall see, can simultaneously arise in policy debates in interrelated ways. We also show that failure to be attentive to these value disagreements can hinder discourse about the relevant science, as well as stall or negatively influence policy decisions. Encouraging people to accept particular policies or participate in certain actions will thus require engaging stakeholders in a more open and critical discussion about values.

On What Has Value

Policy decisions that are appropriately informed by scientific evidence are also grounded in particular judgments about what is valuable. As indicated earlier, most would agree that evidence that a substance is hazardous to human health should lead to policies that regulate such a substance. This is so because most people value human health. Many of the values that underlie public policies are, however, not shared ones. In these cases, people can reasonably disagree about the appropriateness of such policies while agreeing that the existing scientific evidence is nonetheless consistent with the policy in dispute.

For instance, consider conservation efforts in the United States regarding the gray wolf. Gray wolves were listed as endangered in 1974, shortly after the creation of the Endangered Species Act. There was little dispute at the time that the wolves met the scientific criteria for being considered endangered: the number of wolves and breeding pairs were unsustainable and they occupied less than 5% of their historical range.[16] The listing of the species as endangered resulted in prohibitions against the killing and hunting of wolves. It also imposed requirements on the states to develop plans to recover wolf populations in areas such as Yellowstone National Park, where the wolves had once flourished. During the 1980s and 1990s, when scientists developed plans for recovery and reintroduction, however, there was intense controversy.[17] Some supported wolf reintroduction in target areas, such as Yellowstone, because the wolf had been a symbol of the Park and part of its appeal to tourists.[18] Hunters and ranchers in Idaho, Montana, and Wyoming, however, expressed concerns that wolves would pose a threat to livestock when they inevitably traveled outside the park.[19] They argued that the wolves had long been virtually extinct in the area and that reintroducing a species merely because it had once lived there put livestock at unnecessary risk. Conservation biologists alternatively argued that wolves played a vital role in the stability and integrity of the ecosystem. Without them, elk and coyote populations were out of control, which also put a stress on fox (as prey for coyotes) and on certain kinds of vegetation (which served as food for elk).[20] Thus, biologists supported targeted reintroduction plans where reintroduction of the species could benefit unique ecosystems. Some environmentalists argued that recovery plans did not go far enough precisely because wolves should be protected everywhere they had historically inhabited and not just within arbitrary park boundaries.[21]

All parties to these disputes agreed with the scientific evidence that the numbers of gray wolves were unsustainable and that they were near extinction. They nonetheless disagreed about reintroduction and recovery policies because they had conflicting views of whether and why gray wolves had value. Conservation biologists believed that wolves had instrumental value in virtue of the necessary role they seemed to play in the stability and integrity of the Yellowstone ecosystem—something they believed had intrinsic value. Environmentalists who thought that wolves themselves had intrinsic value thought that they should be protected everywhere. Those primarily concerned with tourism were interested in protecting wolves in the park, regardless of whether or not they contributed to the flourishing of the ecosystem, but were less concerned with restoring wolves in other areas of the United States that wolves had once populated. Ranchers and hunters did not support wolf reintroduction because they did not see wolves as valuable but, rather, as competition for something else they *did* find valuable—namely, cattle ranching or hunting elk. Indeed, even after gray wolves were reintroduced into Yellowstone Park in 1995, these disagreements about values continued to manifest in subsequent policy debates about whether to take wolves off the endangered species list and about whether to relist them as endangered once they had been delisted in 2008 in Montana and Idaho.[22]

Arguably, disagreements about what has value also underlie some of the debate about particular climate change policies. One might accept the scientific consensus that average global temperature will increase 2° to 4°C over the next 100 years owing to human activities and that this will lead to higher sea levels, increased droughts, and so forth in the future. But, whether one supports regulations on such emissions depends on, among other things, whether one believes that we have moral obligations to future generations— obligations that can impose burdens on present individuals in order to avoid harms to future ones. If one thinks that no such obligations exist, then one may be less inclined to support policies geared to curbing greenhouse gas emissions. Those supporting policies that restrict CO_2 emissions might do so because they believe these obligations do exist and are likely to take grounds for denying obligations to future generations as morally arbitrary. Proponents and opponents of CO_2 restriction policies can thus disagree about the appropriateness of such policies, but they do so because of their views regarding what has value or what values our policies should aim to protect.

If, indeed, disputes over climate change and wolf recovery plans are disagreements about what has value, this has important implications for how we can better advance policy discussions. If the disagreement is about what is worth protecting, then appealing to evidence that, for example, average global temperature will increase does not address such concerns. It necessitates having a debate about the values at stake. Identifying these values is important because, as mentioned, some of them might be problematic. Indeed, the claim that we have no obligations to future generations is likely made on morally arbitrary grounds and is difficult to defend.

Disagreements about what has value also have implications for the sort of empirical evidence that can be relevant for addressing certain concerns.[23] For instance, if the reason some people reject climate change policies is that they do not believe we have obligations to future generations, then convincing them to accept emissions restrictions may require producing empirical evidence about how climate change will impact present people or their interests. That is, although some disputes about policies do not result from rejection of relevant empirical evidence, they can have implications for the sort of scientific evidence that is needed to move toward a public consensus on policy. For example, currently, scientists are devoting a significant amount of resources to developing global circulation models (GCMs), which are very good at predicting average increases in global temperature, but that tell us little about how particular regions or communities are likely to be impacted.[24] If what stakeholders value is protecting their homes, local food supply, water availability, or other such goods that might be impacted, regional climate models (RCMs) are far more useful than GCMs in generating the right kind of information, even though RCMs may be less useful for helping us understand the climate system as a whole. GCMs may be more appropriate if we are concerned about the average global temperature and broad environmental impacts that might result, and less concerned with the specific impacts on particular communities. Different assumptions about what is valuable have different implications for the kind of evidence that is relevant to support particular policies even if one accepts evidence about the existence of anthropogenic climate change. When people appear to disregard science, it may be that the existing scientific evidence simply does not speak to the empirical question that is relevant, given their values. Thus, a recognition that the disagreement is about values and an exploration of the particular values at stake in policy can help scientists produce the right sort of evidence that will be more responsive to some people's particular concerns.

On How to Interpret Values

Some public policies attempt to protect goods that are valued by most people. For instance, policies that regulate carcinogenic substances attempt to protect a good valued by most human beings—that is, health. In some cases, however, although a policy aims to preserve a good that is broadly valued, the good in question can have various interpretations. In these cases, disagreements about whether the policies in question should or should not be implemented, or how one should go about doing so, can arise because of differences on how to interpret those goods.

In 2002, the Bush administration introduced the Healthy Forests Initiative as a means to enhance forest health on public lands. Congress passed parts of the initiative in the Healthy Forests Restoration Act (HFRA) of 2003. The purpose of the HFRA was to reduce fuel loads on public lands that were at high risk for catastrophic wildfires.[25] But the HFRA, which involved strategies to actively manage forests including mechanical trimming, strategic logging, and prescribed burning, was thoroughly criticized by opponents as a deceptively named policy that ignored ecological science in order to promote the interests of the logging industry.[26] Proponents of the act, however, characterized their critics as extreme environmentalists who ignored evidence that failure to actively manage forests had led to a buildup of fuel load that resulted in more intense and increasingly destructive forest fires. Both sides were concerned with protecting forests, but they had different conceptions of what "forest health" involved,[27] which led them to disagree about what policies were needed. Those who opposed the HFRA tended to understand forest health as involving the natural functioning of an ecosystem without interference from humans. They thus viewed fires as playing a natural—that is, healthy—role in ecosystems, and attributed problems to human activities that led to more devastating fires, such as climate change.[28] Proponents, on the other hand, tended to view humans as an integral part of ecosystems. They saw healthy forests as those that enabled humans to live in and engage in sustained economic activities.[29] For them, achieving forest health required protecting homes and preserving a certain level of logging. Of course, both of these conceptions of what forest health involves may be overly simplistic, but moving the debate forward requires identifying and evaluating the competing interpretations of these values rather than debating the empirical evidence.

Some disputes about climate change adaptation policies also stem from disagreements about how to understand shared values. Climate change adaptation policies seek to change infrastructure and practices so as to minimize the harmful impacts of climate change. Virtually everyone agrees that adoption of aggressive policies to mitigate global warming would be insufficient to prevent harmful effects altogether and therefore some adaptation will be required.[30] But different stakeholders have conflicting views about what this harm minimization should involve when defending adaptation policies and thus they support different policy strategies. For instance, one might think that harm minimization must presuppose that economic development will not be reduced. Those who interpret "minimizing harm" in this way thus support policies that aim to adapt to the sorts of changes that would occur in the climate system if emissions were not simultaneously curbed. Alternatively, others take harm minimization to presuppose that constraints on development will also be put in place. For them, appropriate policies thus require adapting to the changes that are likely to be produced in a context where we also are engaging in aggressive strategies to prevent global warming. Similarly, there are disagreements about whether to understand harm minimization as primarily involving concern for physical goods such as life or property,[31] or also a concern for people's ways of life.[32] In the first case, one might support adaptation policies that help people relocate farther inland, so as to deal with the effects of raising sea levels. Such policies would remove people from vulnerable areas and prevent further loss of life or property. Those who believe people's ways of life matter, however, see those policies as inadequate because relocation often requires that people leave areas that have shaped their economies, skills, language, and cultural traditions. They are thus more likely to support policies that enable communities to stay in the same geographical places or that at least impose constraints on attempts to relocate populations.[33] Although proponents and opponents often present these debates as involving confusion about the scientific evidence on how grave the impacts of climate change and rising sea levels will be, this need not be the case. Different conceptions of what it means to "minimize harm" can give rise to different kinds of policy approaches.

As with the case of disagreements regarding what is valuable, attention to the role that different interpretations of shared values can have in policy decisions is essential to moving policy debates forward in fruitful ways. Neglecting the role of values can further stall needed policies because stakeholders simply talk past one another. Furthermore, diverse conceptualizations of particular values can lead people to conceive of policy aims in alternative ways, which in turn will lead them to assess the relevance of existing evidence differently. Evidence that increased logging will serve to protect homes, for instance, will do little to persuade those who have an ecological conception of forest health about the pertinence of promoting logging. Similarly, evidence about what climate change adaptations will be required to minimize harm in India if the United States drastically decreases emissions will not be compelling to those who think the United States should not be expected to minimize harm by decreasing CO_2 emissions.

On How to Weigh Competing Values

Another reason why people might disagree about policy interventions even when they do not question the existing scientific evidence has to do with the different ways in which people might weigh various shared goods. Disputes about home birth policies illustrate how this can occur. For the past two decades, there has been intense debate about whether home birth for low-risk women is a safe alternative to hospital birth. Both sides of the debate have taken firm positions. The American, Australian, and New Zealand colleges of obstetricians and gynecologists are opposed to home birth.[34] The American Medical Association (AMA), for instance, has adopted a resolution strongly opposing home births for any woman, arguing that the safest setting for labor and delivery is in the hospital or a birthing center within a hospital complex.[35] At the same time, organizations such as the U.K.'s Royal College of Obstetrics and Gynecology, and Royal College of Midwives, as well as the American Public Health Association and the American, Australian, New Zealand, and Canadian colleges of midwives, support the option of home birth for low-risk women.[36]

Although disputes over home-birth policies are often presented as mainly involving disagreements about the existing scientific evidence on the relative safety of home birth,[37] both sides agree on the challenges and limitations of particular home-birth safety studies.[38] They agree, to a large extent, on the current evidence about risks related to home birth. In addition, both sides agree on the values at stake in this policy decision and how to interpret them. Both agree that mother and infant safety are important goods that ought to be promoted by health policies and recommendations. Furthermore, neither side denies that patient autonomy is an important consideration in regulating health decisions, such as where to give birth.

Still, disagreements exist about how much weight to give to each of these values, or which value should have priority when they conflict. Those who oppose policies that allow home birth take patient safety to be of the upmost importance. They point out that even in cases of low-risk pregnancies, serious intrapartum complications may arise without much warning.[39] Given such possibility, they believe that immediate access to expertise and interventions may be life-saving for the mother and the newborn, and may reduce the likelihood of an adverse outcome.[40] In other words, they assume that it is "better to be safe than sorry." Hence, while opponents of home birth might value autonomy, they seem to give greater weight to safety.

Supporters of home-birth programs, on the other hand, seem to give greater priority to respect for patient autonomy. They believe that in countries where infant and maternal morbidity and mortality are already very low and where appropriate emergency options exist, we should see childbirth as an activity that can enhance family bonds rather than as an inherently risky activity.[41] Some women who chose home birth do so because they think a very small risk of a catastrophic consequence is acceptable in a context where the perceived benefits of a home birth, such as having a familiar environment, more control over the birthing process, the ability to choose, and the decreased risk of obstetric interventions, are significant.[42] Thus, the disagreement arises because of different assumptions about how competing goods such as safety and autonomy ought to be weighed against each other, rather than about the strength of the scientific evidence or its relevance for the safety of home birth.

Refusal to vaccinate children can also result from disagreements about how to weigh and prioritize various values at stake, rather than only from confusion about the existing evidence or its strength. Proponents of pediatric vaccines maintain that protecting public health is important and ought to guide public-health policies.[43] Concerned parents may well agree that public health is important and valuable, but they can also have another value at stake: the health and well-being of their particular child.[44] While policymakers have an obligation to give more weight to the interests of society as a whole, parents may think that they should act in ways that primarily promote their own child's best interests. Prioritizing their children's individual well-being will be even more likely in a context where parents have concerns about the degree of attention institutions are paying to the known, even if rare, complications of childhood vaccines.

The different weights that proponents and opponents of vaccination policies give to conflicting goods also have implications for the kind of evidence that would be relevant to advance the debate. Insofar as parents are reluctant to vaccinate their children because of concerns about their own children, they are unlikely to be moved by evidence that vaccines are relatively safe and effective in achieving herd immunity and protecting public health. It need not be that such evidence is irrelevant to those who resist vaccinating their children. It is that the priority they give to the small, but real risks to their children's health necessitates a different type of evidence— evidence that vaccination will not put their child at an unreasonable risk. What these parents want to know is not only whether vaccines are generally safe and effective in achieving herd immunity against a particular disease but what evidence there is regarding the safety of vaccines for their own children. Understanding the values motivating parents helps reveal the sort of evidence necessary for addressing their concerns.[45]

Debates about climate change policies can also result from divergences in judgments about how to weigh competing values. Recent polling shows that a majority of people in the United States now believe that anthropogenic climate change is occurring and that it is likely to have serious consequences.[46] Nonetheless, they disagree about how serious the consequences of climate change will be for them personally, the extent to which climate change mitigation should be a political priority, and whether the benefits of emission-reduction policies would outweigh the substantial perceived economic costs.[47] That is, even when people agree that anthropogenic climate change is occurring, they disagree about how to weigh the competing goods of different policy options. Even among those who agree that the interests of future generations ought to be considered, for example, there may be disagreement about whether the interests of future people deserve the same weight as those of current individuals, whose prospects and opportunities are likely to be significantly limited if countries adopt particular policies.[48] Similarly, people can question whether we should prioritize funding for any climate change policies when there are other, more immediate social problems, such as war, poverty, and problems with education.[49]

Indeed, positions on climate policy—whether those policies should be prioritized or what policies countries should implement—seem to strongly correlate with certain ideological and political value systems.[50] Those who tend to place great weight on free markets and individual liberty view restrictions aimed at mitigation of harms as the wrong approach to concerns about climate change. They tend to be concerned that regulatory policies will come at the unacceptable cost of hindering important economic benefits. But if resistance to certain policy solutions related to climate change is the result of disagreements about how to weigh the competing values at stake, then insisting on the strength of the scientific evidence or on people's lack of knowledge is unlikely to advance the debates. Acknowledging the values at stake in public policy and engaging stakeholders in a discussion about their values may be more effective in moving people toward support for particular policies.

On How Best to Promote Policy Goals

A further source of disagreement regarding what policies to adopt stems from reasonable disagreements about what the best mechanisms are for promoting particular policy goals. Take again the debate about GMPs policies. A recent report about the state of food insecurity in the world estimates that around one in eight people in the world are suffering from chronic hunger,[51] failing to regularly obtain enough food to sustain an active life. Wide-scale use of technologies for genetic modification are often touted as a necessary strategy to address global hunger and food insecurity.[52] Critics of policies that restrict GMPs point to evidence that genetically modified foods, for instance, are safe for human consumption and that we will need the genetic engineering of crops to increase food production particularly in areas such as Africa, which present agricultural challenges.[53] While questions regarding the long-term safety of GMPs remain,[54] opponents of their wide-scale use need not reject evidence that such products are safe. Critics can also share the aim of addressing world hunger and food insecurities. However, many remain unconvinced that wide-scale implementation of genetic engineering technologies are the best means for addressing this problem.[55] This is so not only because evidence indicates that introduction of genetically modified crops in places like the United States has failed to increase yields,[56] but also because problems related to global hunger and food scarcity are complex ones that arguably result not simply from lack of food production but also from poverty, waste, war and conflicts, natural disasters, and distribution problems. Moreover, critics of policies promoting wide-scale use of GMPs can be concerned not only with issues of safety but also with the social and political consequences that extensive use of GMPs might have in the global context.[57] Corporations, such as Monsanto, which have invested significant funds in the research and development of genetically modified seeds, protect their investment through contracts with agricultural growers. Such contracts severely restrict seed use (preventing farmers from saving seeds and/or reusing them), control the context within which disputes may be settled, and limit the liability of the company. Many are thus concerned that these practices will exacerbate social and economic inequalities, as resource-poor countries are forced to purchase seeds or to continue to rely on the Global North for food production. Insofar as this is the case, the wide-scale production of GMPs might in fact be able to increase food resources, but not without also consolidating unequal power dynamics between the Global North and the Global South. Disputes about the appropriateness of policies that encourage such production can thus stem from disagreements about value judgments regarding what might be the best ways to achieve a goal that is shared, such as ending world hunger and food insecurity.

Similar disagreements exist about what mechanisms are best for protecting against catastrophic environmental impacts of climate change. Some of those concerns, as we have seen, are related to differences regarding what kinds of sacrifices are justified or how much weight we ought to give to competing goods. However, even among those who might well agree that we should prioritize the prevention of harmful environmental impacts stemming from climate change, there are still disagreements about how best to do so. For instance, in order to address the negative environmental impacts of certain agricultural practices, some have proposed a tax on foods such as beef products, the production of which produces a significant amount of greenhouse gases.[58] Agriculture is one of the primary drivers of climate change and much of it results from the use of seeds to feed livestock.[59] Proponents of a carbon tax on beef have argued that an increase in its price could curb beef consumption and thus reduce harmful environmental impacts. Nonetheless, one might be reluctant to support such policy because of social justice concerns, even when one might be convinced by the evidence that large-scale beef production is harmful to the environment. Such taxes have a disproportionate impact on poorer consumers, many of whom often bear less responsibility for problems related to climate change. There are, of course, a variety of other policy options that could promote the aim of protecting the environment and even reducing meat consumption. These might include imposing stricter agricultural regulations, taxing beef producers, redirecting subsidies toward nonmeat alternatives, and additional policies that would not penalize consumers. Thus, policy disagreements might involve comparative judgments about whether a particular policy is better than alternatives when trying to achieve a desired policy goal, such as reducing harmful environmental changes.

Conclusion

Science-related policy is crucial for addressing pressing needs in relation to public health, the environment, and other social goods. When people oppose certain policies, or when they refuse to act in ways that protect these goods, this can have devastating consequences. Many have focused on the role that problematic dissent has played in creating public doubt, confusion, and mis-understanding about the relevant science, which in turn has led to resistance to certain policies or individual actions, from policies to mitigate climate change, to vaccinating children, to prohibitions against GMP cultivation. While confusion and doubt about the exiting empirical evidence or about its strength can—and surely does—contribute to these problems, disagreements about values can also play a significant role. Values are necessary in grounding choices about how to act and what policies to endorse. Such disagreements can involve what people take to be valuable, how to interpret shared values, how to weigh conflicting values, and what policies are better for promoting certain valuable goals.

The focus on public confusion about scientific claims and the role of dissent in such confusion has neglected the ways that value disagreements can contribute to stalled policy discussions. This is certainly problematic. It leads scientists to incorrectly view those who disagree with certain policy proposals as necessarily scientifically illiterate and to insist on the strength of the evidence. This strategy is unlikely to be effective. Clearly, if the disagreement is about values rather than simply about the evidence, focusing on the evidence will do little to move the debate forward. It also gives incentive to policymakers and others who oppose certain policies to emphasize uncertainty about the science and to demand the gathering of more evidence before the approval of policies. When people think that the only way they can challenge policies is to challenge the science, they will be more likely to do so. If it is clear that the debate is about differences in values rather than about the existing evidence, the argument that "more research is needed" would lose strength. Moreover, ignoring the role of values in policy disagreements prevents their evaluation. At least in some cases, the values at stake may be highly problematic, and thus difficult to defend publicly. Failure to attend to values will leave them concealed and unchallenged. Disregarding the role of values in policy can also hide the need to generate the kind of empirical evidence that would be relevant to ground certain policies or to address stakeholder concerns. Policies that aim to protect against some harms make assumptions about what is valuable and worth protecting. Alternative assumptions about what has value or how to interpret those values may call for different kinds of empirical evidence. In climate change, for example, what evidence is needed for developing effective adaptation policies depends on what it is that we are trying to adapt *to*, but this in turn also has implications regarding the data that we need to collect and the kinds of models that are important. Similarly, if people refuse to vaccinate their children because they are concerned about risks to their particular child, then evidence about the extent to which such vaccines are safe and effective for achieving herd immunity will be unpersuasive. Consequently, attention to values related to policies can help produce science that is more policy-relevant.

Concerns about dissent have been largely motivated by problems such as delayed policies or failures to adhere to policy recommendations. But addressing these problems requires reframing debates to discuss values rather than to focus exclusive on debating the science. If disagreements about policies is caused by disagreements about values, then no amount of scientific evidence—even evidence that laypersons find reliable—will resolve such disagreements. Making clear the ways that values play a role in policy decisions and clarifying what values are at stake in particular cases may go a long way toward advancing policy debates in fruitful ways and producing scientific evidence that is relevant to policy aims.

11

Lessons Learned and New Directions

In 2013, the Intergovernmental Panel on Climate Change (IPCC) published its *Fifth Assessment Report*, containing its strongest language to date about the threats of anthropogenic global warming. It states that it is now extremely likely that most of the warming over the past fifty years is due to human activities.[1] The report also significantly increased estimates for rising sea levels and concluded that certain impacts, such as an absence of ice in the Arctic Ocean, are happening more quickly than scientists previously predicted. The scientific community overwhelmingly agrees with these claims.[2]

Nonetheless, as we have seen, climate policy, particularly in the United States, has been slow to change and there is significant public resistance both to accepting the consensus view and to changing behavior or supporting regulatory policies. Indeed, members of the Trump administration expressly tasked with developing and enforcing climate change policies and agreements have publicly expressed doubt that climate change is caused by human activity and that mitigation or adaptation policies are warranted. The Environmental Protection Agency (EPA) is responsible for writing and enforcing environmental regulations, but EPA Administrator Scott Pruitt has expressed open skepticism about the degree of scientific agreement regarding climate change and its connection to the actions of humankind, and has urged continuous debate in classrooms, public forums, and the halls of Congress.[3] Secretary of the Interior Ryan Zinke, who is in charge of managing millions of square miles of public lands, concedes that climate change is not a hoax, but has declared that is not proven, and has also voiced doubts about the existence of a consensus.[4] Secretary of Energy Rick Perry, who is charged of, among other things, the National Renewable Energy Laboratory, has declared that concerns about anthropogenic climate change are "all one contrived phony mess."[5]

The position of these high-ranking federal officials, who are charged directly or indirectly with protecting environmental and public health from climate change damages, is echoed by a significant portion of the U.S. population, as well as by citizens in other countries.[6] This resistance is difficult to understand when both the scientific facts seem clear and the gravity of the potential impacts call for the urgent adoption of aggressive mitigation and adaptation policies. Moreover, as we have discussed, this disconnect is neither a new phenomenon, as debates on tobacco harms show,[7] nor a phenomenon limited to climate change, as current disputes regarding the safety of pediatric vaccines and genetically modified products (GMPs) demonstrate.

Many have blamed this disconnect on the existence and dissemination of problematic dissenting views that have confused the public about the state of the science and the degree of consensus in the scientific community.[8] As a result, they have tended to adopt a two-pronged approach to addressing the problem: (1) identifying problematic features that typically characterize such dissent in order to discredit dissenters and limit the publicity of dissenting views;[9] and (2) promoting public awareness that a scientific consensus does in fact exist about various controversial policy-relevant science.[10] The aim of these strategies, which rely on the "deficit model" of the public understanding of science, is to correct laypeople's false beliefs so as to arrive at an agreement among the public about policies and behaviors that are urgently needed.

To be clear, we agree that failing to take action on issues such as climate change, or refusing to vaccinate one's children because of fears of a link with autism, is problematic. It seems to us that the scientific evidence for a variety of policy-relevant claims—such as that anthropogenic climate change is occurring and will have devastating effects, that there is no link between the MMR vaccine and autism, or that HIV causes AIDS—is overwhelming. While the history of science demonstrates that it is always possible a scientific consensus is mistaken, we believe that the evidence for these claims is substantial. And, we should always use our best scientific knowledge to inform science-relevant public policy. We therefore have no difficulties admitting that many of the dissenting views discussed in this book include false and unsubstantiated claims, seriously specious reasoning, flawed methodologies, mischaracterization of the existing scientific evidence, or exaggeration of uncertainties. Scientific scrutiny is essential to expose many of these problems, but philosophers of science also have an important role to play in pointing out the various ways in which instances of dissent—and consensus—can be epistemically flawed.[11] Indeed, many of the authors discussed in this book have done excellent work in exposing dissent that is scientifically flawed,[12] and pointing out the ways in which various private entities have spent enormous amounts of money to deliberately create doubt and confuse the public about the extent and reliability of evidence supporting certain scientific claims so as to stall regulatory policies.[13] These agreements notwithstanding, we have argued in this book that limiting attention to the role that problematic dissent plays in undermining support for particular policies and actions is both misguided and dangerous and that it would serve us well to reframe the problem in ways that are likely to be more fruitful in trying to address the problems that can stem from such dissent.

Why a Focus on NID Is Misguided

Conceptually, it seems clear that NID—that is, dissent that fails to advance or hinders rather than promotes the aims of knowledge production—can exist and that it can have adverse epistemic and social consequences. However, we have also seen that specific criteria to reliably identify NID are difficult to find.[14] We have critically evaluated three potential ways to identify such dissent: (1) exploring the presence of bad-faith motives; (2) determining whether there has been a failure to play by the rules; and (3) examining the distribution of risks. We have concluded that all three are unsuccessful. Each of these criteria either risks identifying as normatively inappropriate cases of dissent that are epistemically and socially beneficial or fails to capture dissent that many take to be paradigmatic cases of NID, or both.

Climate skeptics, creationists, vaccine refusers, and AIDS denialists no doubt have a variety of motives, and it is not clear that in all cases they provide their dissenting views in bad faith or with intentions that are contrary to epistemic aims. As we argued in chapter 3, motives are very difficult to ascertain. Worse still, even when dissent *is* offered with nefarious motives, it can still be epistemically beneficial, and one can offer epistemically valuable dissent under motives that are suspect. Thus, appealing to dissenters' intentions cannot reliably identify NID.

Similarly, determining whether particular instances of dissent fail to play by the rules can be challenging. In chapter 4 we examined several possible ways in which dissent could fail to conform to the norms that govern transformative criticism, including uptake, shared standards, and expertise. If one conceptualizes these norms in thick, robust ways, then climate change and vaccine skepticism or intelligent design theory would most certainly count as normatively inappropriate. One can make a strong case that at least some of the claims put forth by these dissenters violate shared standards because they employ unorthodox methodologies, background assumptions, and statistical analyses.[15] In addition, many climate and vaccine skeptics lack expertise, when one takes expertise to require a Ph.D. in a particular relevant scientific discipline or a track record of peer-reviewed publications. However, this thick interpretation of the norms for transformative criticism would rule as normatively inappropriate a significant amount of dissent that, by all accounts, could be epistemically and socially valuable. Dissent regarding the safety of GMPs, for instance, is likely to be captured under this thick interpretation,[16] but whether a scientific consensus exists regarding GMPs safety is highly contested.[17] Dissent that originates from those using different disciplinary approaches, methodologies, or background assumptions, and dissent that comes from "outsiders" who may have local knowledge relevant to understanding some scientific phenomena could also be branded as normatively inappropriate. Indeed, in the case of climate change, layperson stakeholders have arguably provided valuable criticisms of climate models and their limitations, particularly insofar as such models have produced data that reflect interests of the Global North while neglecting features of the climate system that may be important to developing socially just climate policies.[18] Clearly, a thick interpretation of the norms of transformative criticism threatens valuable epistemic pluralism.

Alternatively, one could interpret norms such as shared standards or expertise more thinly. Doing so would reduce the risk of excluding epistemically valuable disagreements about methodologies and background assumptions, and would allow participants with various types of valuable expertise to contribute to scientific debates. However, it is not clear whether instances of dissent that many take to be NID would violate these norms so understood. Therefore, using norms of transformative criticism as criteria to identify NID also fails to reliably identify such dissent.

This is not to say that we have no way of knowing or evaluating whether claims made by climate skeptics, vaccine refusers, intelligent-design theorists, or AIDS denialists are false or unjustified. We can and we actually do. But recall from chapter 2 that NID is not just dissent that is simply erroneous. Even dissent that involves false claims can have epistemic and social benefits. Rather, NID is dissent that fails to promote or that hinders scientific progress, and dissent that turns out to be mistaken might be quite epistemically valuable nonetheless. Our point has been to show that using norms of transformative criticism as criteria to identify NID cannot help us make this important distinction.

Finally, we considered whether some dissent qualifies as NID because the risks of such dissent are grave and are distributed unfairly. Falsely accepting a dissenting view can have severe social consequences, and in some cases the potential costs could disproportionately impact the public. At the same time, not uncommonly, those who benefit from falsely rejecting the consensus view are also those who stand to gain financially from a lack of regulatory policies. Yet, as argued in chapter 5, assessing the risks of instances of dissent may also prove complex. Both falsely rejecting and falsely accepting the consensus on climate change, for instance, can have severe effects on various stakeholders. Furthermore, predicting risks and determining on whom the risks fall and how much weight to give to different types of possible harms and benefits is challenging, and reasonable people might disagree about such judgments. After all, neither the public nor the producers are monolithic groups with universally shared interests.

Clearly, coming up with criteria to reliably identify NID is more difficult than it might initially appear. Thus, insofar as the goal is to develop strategies that might address problems such as confusion or stalled policies, this approach would be ineffective. This does not imply, however, that the criteria discussed here could not be useful for some other purposes. For example, the concept of bad-faith motives may be important to understanding and addressing epistemic vices.[19] Considerations such as uptake and shared standards can be critical criteria for assessing the objectivity of scientific communities.[20] Conceptions of expertise are relevant, for example, for deciding who should serve on science advisory boards. Assessing risks may be important to determining how much evidence is needed for accepting hypotheses or in helping us think about the epistemic duties that dissenters have. Our claim, however, is that these criteria are not effective in reliably identifying NID.

Yet focusing on dissent is also misguided because it can obscure other factors that affect why people may be resistant to particular policies or behaviors. As we argued in chapter 10, opposition to certain policies and actions can stem from disagreements about values that necessarily underlie policy decisions, rather than rejection of or confusion about empirical claims.[21] People may disagree about what has value, how to interpret broadly shared values, how to weigh or prioritize values when they conflict, and how best to promote particular policy goals, all of which may have implications for whether one believes that particular policies are justified.

Consider, for instance, disputes about climate change. An assumption, implicit and explicit, that underlies concerns about NID is that laypersons are confused about or misunderstand relevant science, and that this leads them to reject public policies that address global warming. However, as we have seen, empirical evidence suggests that those who have a strong sense of social justice are more likely to support climate change mitigation policies than those who give more weight to values such as wealth.[22] Egalitarian worldviews also strongly predict support for climate change policies even when they might be costly.[23] Perceived costs or threats to other values have also been a factor in public rejection of particular climate policies. For example, some people state concerns that policies will negatively impact jobs or the economy as a reason for why they do not support climate policies.[24] Policy support also depends on value judgments about who has responsibility for acting.[25] Thus, it is unhelpful to assume that resistance to particular policies or behavioral changes are primarily a consequence of NID.

If disagreements about policies are the result of differences in values, then the deficit model of the public of understanding of science is false, or at least incomplete. Unfortunately, misdiagnosing the problem can also lead to misguided solutions. In chapter 6 we argued that some of these strategies, such as emphasizing the agreement that exists among scientists, are unlikely to be particularly effective, and at least some sectors of the public can perceive them as insulting. This is not to say that science communication and education have no role in democratic societies. They certainly do. Our claim is that insisting that the problem is one primarily involving misunderstandings resulting from wrong-headed dissent will do little to improve policy debates.

Fruitfully moving policy discussions forward requires refocusing the discussion in ways that identify and engage with the values people hold. To begin with, inattentiveness to the relevance of value disagreements at stake can undermine communication efforts regarding scientific evidence and lead people to talk past one another. For instance, some who are less enthusiastic about restrictions on greenhouse gas emissions may want evidence not only that anthropogenic climate change is occurring and that it will have global impacts but also information about how this will affect them or the things they care about. They may need persuading that the benefits of regulations will outweigh the costs and that those costs will be fairly distributed. For example, a study in the United States and Canada showed that support for restrictions on greenhouse gas emissions increased among people who were told that costs for these regulations would be shared internationally, which addressed some of their concerns.[26] Similarly, addressing parental refusal to vaccinate children requires not just insisting on the relative benefits of vaccines and on their public health benefits but also attending to parents' legitimate concerns regarding the small but existent and serious risks to the health of their children, or what appears to be an inadequate monitoring system.[27]

Moreover, inattention to the necessary function that values play in policymaking allows values that are problematic to go uncritically unchecked. Although disagreements about values can be contentious and difficult to solve, ignoring their role would hardly improve things. Bringing values into the open allows for their critical evaluation and, when appropriate, their rejection from grounding any policy decisions.

Why a Focus on NID Is Dangerous

Unfortunately, focusing on problematic dissent is not simply misguided but also dangerous. As also seen in chapter 6, while having criteria for NID might allow one to target dissent in some way, many of the possible strategies that can be used to do so run the risk of exacerbating the very problems they seek to solve, and they may create new ones. As we saw, prohibiting or preventing NID presents practical difficulties at least in democratic nations, and raises serious ethical concerns. Similarly, ignoring NID would permit its unchallenged dissemination, while devoting a lot of resources to responding to it would divert resources away from more fruitful research tasks. Additionally, as the "climategate" incident shows, minimizing opportunities for dissenters may give the public the impression that scientists fear critical examination of their views or that they are hiding something.[28] This, in turn, can further undermine warranted public trust in climate scientists. Emphasizing consensus may be important for persuading laypersons, but it can mask genuine disagreements that are relevant to policy decisions. Moreover, publicizing the fact that a consensus exists can backfire insofar as the public may be skeptical of how scientists achieved such a consensus.[29] Discrediting dissenters by, for example, exposing their financial or political interests can be valuable for a variety of reasons, but risky in a context where more and more scientists have such conflicts. Thus, as we argue in chapter 9, at the very least more needs to be done to address conflicts of interest among scientists and to facilitate well-grounded public trust.

A focus on NID is also dangerous because it keeps policy debates centered exclusively on scientific claims. In doing so, it provides those who wish to thwart policies they find undesirable with a strong incentive to attack those scientific claims. The underlying assumption that consensus is essential to justify policy action provides critics of the policies with very good reasons to conduct and disseminate dissenting views and to call attention to the uncertainties present in scientific research. The focus on NID, then, actually contributes to the problem that such a focus aims to solve. Moreover, by attending to the science, those who oppose unwanted policies are allowed to remain unaccountable for the value judgments that drive their policy preferences. And, as we mentioned earlier, in many cases these value judgments may be problematic or rest on morally arbitrary grounds.

Similarly, in conceding, or appearing to concede, that near unanimous agreement on the science is necessary to justify policymaking, a focus on NID is also dangerous because it constrains what can be appropriate justifications for policies. Significant uncertainties in policy-relevant science are common, and yet decisions to act must be made anyway. Even if there were disagreements among scientists or uncertainties surrounding anthropogenic global warming, for instance, we might still think that the potential risks of doing nothing are far too grave.

Finally, a focus on NID is dangerous because it obscures other features of the context in which science is practiced that can make instances of dissent particularly damaging. In particular, it ignores the obligations that scientific communities have in creating institutional and scientific practices that facilitate and sustain warranted trust, and it neglects the serious costs to science and society that arise from a context where well-grounded public trust in science is precarious. Neglecting the reasons that the public may have for doubting scientists and their claims opens the door for NID to inflict more serious epistemic and social damages.

Where Do We Go From Here?

Given the challenges that both reliably identifying and dealing with NID have, we call for a reframing of the problem. Rather than simply focusing on whether dissent is normatively inappropriate, there should be more attention to the social context in which dissent arises. As we have shown, some social contexts provide NID with a more nurturing environment, making it more likely that it will have more damaging effects. NID can do more damage in a context that fails to maintain warranted epistemic trust. Moreover, NID carries more significance than it should in a context where policy debates are primarily focused on disagreements over scientific claims, instead of attending to the value judgments where disagreements among stakeholders may actually lie.

Belief and acceptance of scientific claims require epistemic trust. When institutional and social practices facilitate and sustain warranted public trust, people will be more likely to find consensus views credible, even in the face of NID. However, when such institutions and practices cast doubt on the trust-worthiness of scientists, NID finds fertile soil. In chapter 7, we explained how exactly this can occur. In a context that is trust undermining, laypersons can become confused because there is no reliable guide regarding whom or what to believe. In such a context, the public and policymakers are more likely to find NID compelling, particularly when it coheres with other background beliefs that form the basis for distrust. Claims that, for instance, climate scientists are fabricating or cherry-picking data find a nurturing setting in a context where people believe that scientists are driven by political agendas that bias their work. In addition, while consensus is often highly publicized, scientists pay less attention to those features that make a consensus particularly reliable. For example, even the IPCC, which is very transparent about the processes it has for the approval of its assessment reports, is notably opaque about the ways in which it reaches consensus when generating and writing those reports.[30] In a context where scientists are not credible, the mere existence of a consensus will do little to reassure people about the integrity of the science.

Attending to structural factors that can facilitate and maintain warranted epistemic trust reveals strategies better suited for minimizing the potential negative impacts of NID regardless of whether particular instances of dissent are, in fact, normatively inappropriate. As we saw in chapters 8 and 9, various factors can call into question the trustworthiness of scientific communities and thus undermine warranted public trust. People expect that scientific knowledge will improve the common good, rather than for the good of just some stakeholders. In a context where science is increasingly a means of obtaining profit, the common good might sometimes be neglected. Similarly, where financial rewards accompany the conduct of much research, financial conflicts of interest become prevalent for both individual scientists and re-search institutions. While it seems quite relevant to point out that climate change dissent is often funded by powerful economic interests, such as oil companies, doing so might not be very successful in a context where conflicts of interest are rampant among scientists and institutions. That is, laypersons can reasonably infer that no one in the debate is trustworthy and therefore refuse to accept any particular claim or commit to far-reaching policies. Similarly, charges that climate scientists cherry-pick or fabricate data, even when completely unfounded, have more resonance in a context where other cases of scientific misconduct are highly publicized or where policies to detect and penalize misconduct are not particularly effective.

The perception that consensus scientists have a particular political or progressive environmentalist agenda also has the potential to undermine trust in the epistemic integrity of the research. There is concern that scientists will allow their preferences for particular agendas to influence or supplant the evidence for their claims. Laypersons who hold more conservative ideologies are often skeptical of climate scientists and of organizations such as the Union of Concerned Scientists because they take them to have a liberal environmentalist agenda. To address mistrust related to value-driven science, it may be tempting to think that it would be good for scientists to adhere more strictly to a value-free ideal of science, so as to ensure that nonepistemic values do not influence scientific reasoning. But, as we saw in chapter 9, this would be misguided, as such values play important and necessary roles in scientific reasoning. For instance, climate change research aims at trying to protect the things we care about, and what that may be is grounded on value judgments. Research aimed at identifying and measuring the impacts and threats of climate change or at assessing the benefits and risks of various strategies to mitigate or adapt to the changing climate is likely to require a variety of value judgments related to the well-being of humans, nonhuman animals, and the environment. Facilitating trust in this context requires establishing mechanisms to protect the epistemic integrity of the research and to allow values to play only legitimate roles in scientific reasoning. Moreover, people must be aware that such mechanisms exist.

Strategies for promoting trust in the epistemic integrity of the research, however, may fail to address another source of concern for laypersons. Even if values can legitimately operate in science, people can still be concerned about who decides which values to endorse or prioritize. Scientists are not value experts, and their choices are unlikely to be representative of the diverse values that exist among stakeholders. Thus, creating and sustaining warranted public trust also requires the implementation of strategies that promote transparency and stakeholder input in inclusive and representative ways.

Moreover, recognizing the limits of science in the context of policymaking is a second way to reframe the problem of NID in a more successful way. As we argued in chapter 10, some of the problems attributed to NID are the result of disagreements about values, not about facts. Insofar as this is the case, moving policy debates forward in fruitful ways requires engaging in discussions with all relevant parties about the values at stake, rather than the truth of particular scientific claims. Scientific evidence alone does not determine policy and, insofar as values are involved, disagreements about values cannot, and should not, be decided by scientists alone.

While we believe that worries about the trustworthiness of scientists and disagreement about values underlie people's resistance to various policy initiatives, more work is required. Specifically, additional social science research is needed to identify a full range of sources of distrust, determine what particular mechanisms are useful in facilitating trust, and provide information about how to effectively communicate the existence of those mechanisms to the public so as to enhance warranted trust. Moreover, additional work is necessary to determine ways that scientists can collaborate effectively with stakeholders so as to involve them in value judgments related to research. This will be crucial both to ensuring that appropriately justified values guide scientific inquiry and to assuaging stakeholders' legitimate concerns about the influence of contextual values in science. As we noted in chapter 10, climate scientists and social scientists have done a significant amount of work to involve stakeholders in climate impacts and adaptation research.[31] Future research will need to provide additional assessment of these efforts.

NOTES

Chapter 1

1. M. Specter, The denialists: the dangerous attacks on the consensus about H.I.V. and AIDS, *New Yorker*, March 12, 2007. M. Weinel, Primary source knowledge and technical decision-making: Mbeki and the AZT debate, *Studies in History and Philosophy of Science* 38, no. 4 (2007): 748–60. N. Nattrass, AIDS and the scientific governance of medicine in post-apartheid South Africa, *African Affairs* 107, no. 427 (2008): 157–76. N. Simelala et al., A political and social history of HIV in South Africa, *Current HIV/AIDS Reports* 12, no. 2 (2015): 256–61.

2. R. Horton, Mbeki defiant about South African HIV/AIDS strategy, *Lancet* 356, no. 9225 (2000): 225. P. Chigwedere and M. Essex, AIDS denialism and public health practice, *AIDS and Behavior* 14, no. 2 (2010): 237–47.

3. P. Chigwedere et al., Estimating the lost benefits of antiretroviral drug use in South Africa, *Journal of Acquired Immune Deficiency Syndrome* 49, no. 4 (2008): 410–15. Nattrass, AIDS and the scientific governance.

4. C. Hough-Telford et al., Vaccine delays, refusals, and patient dismissals: a survey of pediatricians, *Pediatrics* 138, no. 3 (2016): e20162127. M. S. Majumder et al., Substandard vaccination compliance and the 2015 measles outbreak, *JAMA Pediatrics* 169, no. 5 (2015): 494–95.

5. D. K. Flaherty, The vaccine-autism connection: a public health crisis caused by un-ethical medical practices and fraudulent science, *Annals of Pharmacotherapy* 45, no. 10 (2011): 1302–304.

6. Z. Horne et al., Countering antivaccination attitudes, *Proceedings of the National Academy of Sciences USA*, 112, no. 33 (2015): 10321–24.

7. Majumder et al., Substandard vaccination compliance.

8. N. Watts et al., Health and climate change: policy responses to protect public health, *Lancet* 386, no. 10006 (2015): 1861–914.

9. World Health Organization (WHO), Climate change and health. Fact Sheet. WHO, 2017, http://www.who.int/mediacentre/factsheets/fs266/en/.

10. S. C. Anenberg et al., Global air quality and health co-benefits of mitigating near-term climate change through methane and black carbon emission controls, *Environmental Health Perspectives* 120, no. 6 (2012): 831–39. A. Woodward et al., Climate change and health: on the latest IPCC report, *Lancet* 383, no. 9924 (2014): 1185–89.

11. B. R. McFadden, Examining the gap between science and public opinion about genetically modified food and global warming, *PLOS One* 11, no. 11 (2016): e0166140. M. A. Ranney and D. Clark, Climate change conceptual change: scientific information can transform attitudes, *Topics in Cognitive Science* 8, no. 1 (2016): 49–75. R. E. Dunlap and A. M. McCraight, A widening gap—Republican and Democratic views on climate change, *Environment* 50, no. 5 (2008): 26–35. K. S. Fielding and M. J. Hornsey, A social

identity analysis of climate change and environmental attitudes and behaviors: insights and opportunities, *Frontiers in Psychology* 7, no. 121 (2016): doi: 10.3389/fpsyg.2016.00121. A. Zia and A. M. Todd, Evaluating the effects of ideology on public understanding of climate change science: how to improve communication across ideological divides? *Public Understanding of Science* 19, no. 6 (2010): 743–61. D. M. Kahan, Ideology, motivated reasoning, and cognitive reflection, *Judgment and Decision Making* 8, no. 4 (2013): 407–24.

12. D. Michaels, *Doubt Is Their Product: How Industry's Assault on Science Threatens Your Health* (Oxford and New York: Oxford University Press, 2008). N. Oreskes and E. M. Conway, *Merchants of Doubt: How a Handful of Scientists Obscured the Truth on Issues from Tobacco Smoke to Global Warming* (New York: Bloomsbury, 2010). R. Proctor, *Golden Holocaust: Origins of the Cigarette Catastrophe and the Case for Abolition* (Berkeley: University of California Press, 2011). J. Biddle and A. Leuschner, Climate skepticism and the manufacture of doubt: can dissent in science be epistemically detrimental? *European Journal for Philosophy of Science* 5, no. 3 (2015): 261–78. D. Harker, *Creating Scientific Controversies: Uncertainty and Bias in Science and Society* (Cambridge: Cambridge University Press, 2015).

13. P. Duesberg and D. Rasnick, The AIDS dilemma: drug diseases blamed on a passenger virus, *Genetica* 104, no. 2 (1998): 85–132. P. Duesberg, C. Koehnlein, and D. Rasnick, The chemical bases of the various AIDS epidemics: recreational drugs, anti-viral chemotherapy and malnutrition, *Journal of Biosciences* 28, no. 4 (2003): 383–412. P. H. Duesberg et al., AIDS since 1984: no evidence for a new, viral epidemic—not even in Africa, *Italian Journal of Anatomy and Embryology* 116, no. 2 (2011): 73–92.

14. Chigwedere, Estimating the lost benefits. Nattrass, AIDS and the scientific governance.

15. A. Wakefield et al., Ileal-lymphoid-modular hyperplasia, non-specific colitis, and pervasive developmental disorder in children [retracted], *Lancet* 351, no. 9103 (1998): 637–41. A. J Wakefield and S. M. Montgomery, Measles, mumps, rubella vaccine: through a glass, darkly, *Adverse Drug React Toxicology Review* 19, no. 4 (2000): 265–83; discussion 284–92.

16. Majumder et al., Substandard vaccination compliance. Horne et al., Countering antivaccination attitudes. Flaherty, The vaccine-autism connection. M. J. Smith et al., Media coverage of the measles-mumps-rubella vaccine and autism controversy and its relationship to MMR immunization rates in the United States, *Pediatrics* 121, no. 4 (2008): e836–43. E. Dubé, M. Vivion, and N. E. MacDonald, Vaccine hesitancy, vaccine refusal and the anti-vaccine movement: influence, impact and implications, *Expert Review of Vaccines* 14, no. 1 (2015): 99–117.

17. F. Godlee, J. Smith, and H. Marcovitch, Wakefield's article linking MMR vaccine and autism was fraudulent, *British Medical Journal* 342 (2011): c7452.

18. Flaherty, The vaccine-autism connection.

19. N. Oreskes and E. M. Conway, Defeating the merchants of doubt, *Nature* 465, no. 7199 (2010): 686–87. Oreskes and Conway, *Merchants of Doubt*.

20. H. Schmid-Petri et al., A changing climate of skepticism: the factors shaping climate change coverage in the U.S. press, *Public Understanding of Science* 25, no. 4 (2017): 498–513.

21. D. J. Koehler, Can journalistic "false balance" distort public perception of consensus in expert opinion? *Journal of Experimental Psychology Applications* 22, no. 1 (2016): 24–38.

22. A. Leiserowitz et al., *Climate Change in the American Mind: March 2016* (New Haven, CT: Yale University Press, 2014). S. L. van der Linden et al., The scientific consensus on climate

change as a gateway belief: experimental evidence, *PLOS One* 10, no. 2 (2015): e0118489. J. Shi, V. Visschers, and M. Siegrist, Public perception of climate change: the importance of knowledge and cultural worldviews, *Risk Analysis* 35, no. 12 (2015): 2183–201. M. Aklin and J. Urpelainen, Perceptions of scientific dissent undermine public support for environmental policy, *Environmental Science and Policy* 38 (2014): 173–77.

23. Biddle and Leuschner, Climate skepticism.

24. H. E. Longino, *The Fate of Knowledge* (Princeton, NJ: Princeton University Press, 2002). P. Kitcher, *Science in a Democratic Society* (Amherst, NY: Prometheus, 2011).

25. H. M. Collins and R. Evans, *Rethinking Expertise* (Chicago: University of Chicago Press, 2007). E. Selinger and R. P. Crease, *The Philosophy of Expertise* (New York: Columbia University Press, 2006). W. E. Bijker, R. Bal, and R. Hendriks, *The Paradox of Scientific Authority: The Role of Scientific Advice in Democracies* (Cambridge, MA: MIT Press, 2009). S. Jasanoff, *States of Knowledge: The Co-Production of Science and Social Order* (London and New York: Routledge, 2004).

26. T. O. McGarity and W. Wagner, *Bending Science: How Special Interests Corrupt Public Health Research* (Cambridge, MA: Harvard University Press, 2008). Michaels, *Doubt Is Their Product.* Oreskes and Conway, *Merchants of Doubt.* S. B. Omer et al., Vaccine refusal, mandatory immunization, and the risks of vaccine-preventable diseases, *New England Journal of Medicine* 360, no. 19 (2009): 1981–88.

27. McGarity and Wagner, *Bending Science.* Michaels, *Doubt Is Their Product.* Oreskes and Conway, *Merchants of Doubt.* Proctor, *Golden Holocaust.*

28. L. Ceccarelli, Manufactured scientific controversy: science, rhetoric, and public debate, *Rhetoric and Public Affairs* 14, no. 2 (2011): 195–228. Harker, *Creating Scientific Controversies.*

29. Biddle and Leuschner, Climate skepticism. A. Leuschner, Is it appropriate to "target" inappropriate dissent? On the normative consequences of climate skepticism, *Synthese* 195, no. 3 (2018):1255–1271.

30. We talk about "facts" and "values" for simplification purposes. As some of our later chapters will discuss at length, we do not believe that a clear-cut distinction between facts and values exists. With this language we simply want to call attention to the difference between what is usually thought to be a matter primarily of empirical evidence—and where, of course, values play a role—and what might be a matter of value judgment, and where, of course, empirical evidence can be relevant.

31. J. Delbore, Transgenes and transgressions: scientific dissent as heterogeneous practice, *Social Studies of Science* 38, no. 4 (2008): 509–41.

32. Oreskes and Conway, *Merchants of Death.* Harker, *Creating Scientific Controversies.* Biddle and Leuschner, Climate skepticism. Proctor, *Golden Holocaust.*

33. Oreskes and Conway, *Merchants of Death.*

34. Biddle and Leuschner, Climate skepticism.

35. M. Solomon, *Norms of Dissent,* Centre for the Philosophy of Natural and Social Science Contingency and Dissent in Science, Technical report 09/08, 2–21, www. Lse. ac.uk/CPNSS/projects/CoreResearchProjects/ContingencyDissentInScience/DP/SolomonNormsOfDissnt0908Online.pdf.

36. Harker, *Creating Scientific Controversies.*

37. Ideally, criteria for NID would spell out necessary and sufficient conditions so that it can capture all cases of NID while excluding dissent that is epistemically valuable.

However, we recognize that this bar might be unnecessarily high. Determining sufficient conditions for NID would allow us to identify at least some cases of such dissent while protecting that which is epistemically valuable. Hence, successful criteria would pinpoint at least sufficient conditions of NID.

38. We are not suggesting that the contested nature of dissent in these three cases is the same, only that in greater or lesser degree some question the existence of a reliable consensus.

39. Kitcher, *Science in a Democratic Society.* Harker, *Creating Scientific Controversies.*

40. J. Romeis, M. A. McLean, and A. M. Shelton, When bad science makes good headlines: Bt maise and regulatory bans, *Nature Biotechnology* 31, no. 5 (2013): 386–87. A. Nicolia et al., An overview of the last 10 years of genetically engineered crop safety research, *Critical Reviews in Biotechnology* 34, no. 1 (2014): 77–88.

41. M. Cuhra, Review of GMO safety assessment studies: glyphosate residues in Roundup Ready crops is an ignored issue, *Environmental Sciences Europe* 27 (2015): 20. J. A. Magana-Gomex and A. M. C. de la Barca, Risk assessment of genetically modified crops for nutrition and health, *Nutrition Reviews* 67, no. 1 (2009): 1–16. W. Zhang and E. Shi, Do genetically modified crops affect animal reproduction? A review of the ongoing debate, *Animal* 5, no. 7 (2011): 1048–59. D. E. Bushey et al., Characteristics and safety assessment of intractable proteins in genetically modified crops, *Regulatory Toxicology and Pharmacology* 68, no. 2 (2014): 154–70.

42. Kitcher, *Science in a Democratic Society.* Nicolia et al., An overview of the last 10 years. B. M. Maghari and A. M. Ardekano, Genetically modified foods and social concerns, *Avicenna Journal of Medical Biotechnology* 3, no. 3 (2011): 109–17. J. L. Domingo and J. G. Bordonaba, A literature view on the safety assessment of genetically modified plants, *Environment International* 37, no. 4 (2011): 734–42. C. Snell et al., Assessment of the health impact of GM plant diets in long-term and multi-generational animal feeding trials: a literature review, *Food and Chemical Toxicology* 50, nos. 3–4 (2012): 1134–48. A. Dona and I. S. Aranitoyannis, Health risks of genetically modified foods, *Critical Reviews in Food Science and Nutrition* 49, no. 2 (2009): 164–75. H. Lacey, *Values and Objectivity in Science* (Lanham, MD: Rowman & Littlefield, 2005). A. Hilbeck et al., No scientific consensus on GMO safety, *Environmental Sciences Europe* 27, no. 4 (2015), https://doi.org/10.1186/s12302-014-0034-1.

43. Michaels, *Doubt Is Their Product.* Oreskes and Conway, Defeating the merchants of doubt. Oreskes and Conway, *Merchants of Doubt.* N. Oreskes, Beyond the ivory tower: the scientific consensus on climate change. *Science* 306, no. 5702 (2004): 1686. Proctor, *Golden Holocaust.* McGarity and Wagner, *Bending Science.* Leuschner, Is it appropriate.

44. Oreskes, Beyond the ivory tower. Cook et al., Consensus on consensus: a synthesis of consensus estimates on human-caused global warming, *Environmental Research Letters* 11, no. 4 (2016): 048002.

45. J. Cook and P. Jacobs, Scientists are from Mars, laypeople are from Venus: an evidence- based rationale for communicating the consensus on climate, *Reports of the National Center for Science Education* 34, no. 6 (2014): 3.1–3.10.

46. B. Martin, Suppression of dissent in science. In W. R. Freudenburg and T. I. K. Youn, eds., *Research in Social Problems and Public Policy* (Stamford, CT: JAI Press, 1999), 7:105–35.

47. K. Popper, *Conjectures and Refutations: The Growth of Scientific Knowledge* (London: Routledge, 1963). P. Feyerabend, *Against Method: Outline of an Anarchistic*

Theory of Knowledge (Atlantic Highlands, NJ: Humanities Press, 1975). N. Rescher, *Pluralism: Against the Demand for Consensus* (New York: Oxford University Press, 1993). H. E. Longino, *Science as Social Knowledge: Values and Objectivity in Scientific Inquiry* (Princeton, NJ: Princeton University Press, 1990). Longino, *The Fate of Knowledge*. M. Solomon, *Social Empiricism* (Cambridge, MA: MIT Press, 2001). Kitcher, *Science in a Democratic Society*.

48. J. Beatty and A. Moore, Should we aim for consensus? *Episteme: A Journal of Social Epistemology* 7, no. 3 (2010): 198–214.

49. Beatty and Moore, *Should we aim for consensus*.

50. Martin, Suppression of dissent in science. Longino, *Science as Social Knowledge*. Solomon, *Social Empiricism*. Feyerbend, *Against Method*. Popper, *Conjectures and Refutations*. C. Cushman, Intelligent design belongs with Darwin in classrooms; political correctness does not, *U.S. New and World Report*, February 10, 2009, https://www.usnews.com/opinion/articles/2009/02/10/intelligent-design-belongs-with-darwin-in-classrooms-political-correctness-does-not. L. H. Nelson, *Who Knows: From Quine to a Feminist Empiricism* (Philadelphia, PA: Temple University Press, 1990). S. G. Harding, Rethinking standpoint epistemology: what is strong objectivity?, in L. Alcoff and E. Potter, eds., *Feminist Epistemologies* (New York: Routledge, 1993), 49–82. S. G. Harding, *Sciences from Below: Feminisms, Postcolonialities, and Modernities*, Next Wave: New Directions in Women's Studies (Durham, NC: Duke University Press, 2008). E. Anderson, The epistemology of democracy, *Episteme: A Journal of Social Epistemology* 3, no. 1 (2006): 8–22.

51. M. Fricker, *Epistemic Injustice: Power and the Ethics of Knowing* (Oxford and New York: Oxford University Press, 2007).

52. Longino, *Science as Social Knowledge*. McGarity and Wagner, *Bending Science*. Michaels, *Doubt Is Their Product*. Oreskes and Conway, *Merchants of Doubt*. Proctor, *Golden Holocaust*.

53. Harker, *Creating Scientific Controversies*. Oreskes and Conway, *Merchants of Doubt*.

54. T. S. Kuhn, *The Essential Tension: Selected Studies in Scientific Tradition and Change* (Chicago: University of Chicago Press, 1977). Longino, *Science as Social Knowledge*. Longino, *The Fate of Knowledge*. Solomon, *Social Empiricism*. Kitcher, *Science in a Democratic Society*. K. Borgerson, Amending and defending critical contextual empiricism, *European Journal for Philosophy of Science* 1, no. 3 (2011): 435–49.

55. Longino, *Science as Social Knowledge*. Longino, *The Fate of Knowledge*. Kitcher, *Science in a Democratic Society*.

56. Longino, *The Fate of Knowledge*. Harker, *Creating Scientific Controversies*. Cook et al., *Consensus on consensus*.

57. Longino, *Science as Social Knowledge*. Longino, *The Fate of Knowledge*. Nelson, *Who knows*.

58. Biddle and Leuschner, Climate skepticism.

59. Recall that unreliable identification of NID can have serious adverse epistemic and social consequences if the dissent thus identified is in fact legitimate dissent.

60. Oreskes and Conway, *Merchants of Doubt*. Ceccarelli, Manufactured scientific controversy. Harker, *Creating Scientific Controversies*. Biddle and Leuschner, Climate skepticism.

61. Cook et al., Consensus on consensus. S. Van der Linden, Why doctors should convey the medical consensus on vaccine safety, *Evidence-Based Medicine* 21, no. 3 (2016): 119. S.

Van der Linden, C. E. Clarke, and E. W. Maibach, Highlighting consensus among medical scientists increases public support for vaccines: evidence from a randomized experiment, *BMC Public Health* 15 (2015): 1207. Oreskes and Conway, Defeating the merchants of doubt. Oreskes, Beyond the ivory tower.

62. Michaels, *Doubt Is Their Product*. Oreskes and Conway, *Merchants of Doubt*. Harker, *Creating Scientific Controversies*.

63. Cook et al., Consensus on consensus.

64. Longino, *The Fate of Knowledge*. Longino, *Science as Social Knowledge*. Kitcher, *Science in a Democratic Society*.

65. K. Frost-Arnold, Moral trust and scientific collaboration, *Studies in History and Philosophy of Science* 44, no. 3 (2013): 301–10. T. Wilholt, Epistemic trust in science, *British Journal for the Philosophy of Science* 64, no. 2 (2013): 233–53. K. Rolin, Gender and trust in science, *Hypatia* 17, no. 4 (2002): 95–118. J. Hardwig, The role of trust in knowledge, *Journal of Philosophy* 88, no. 12 (1991): 693–708. H. E. Grasswick, Scientific and lay communities: earning epistemic trust through knowledge sharing, *Synthese* 177, no. 3 (2010): 387–409. K. P. Whyte and R. P. Crease, Trust, expertise, and the philosophy of science, *Synthese* 177, no. 3 (2010): 411–25. N. Scheman, Epistemology resuscitated: objectivity and trustworthiness, in N. Tuana and S. Morgen, eds., *Engendering Rationalities* (Albany: State University of New York Press, 2001), 23–52.

66. Grasswick, Scientific and lay communities. M. J. Goldenberg, Public misunderstanding of science? Reframing the problem of vaccine hesitancy, *Perspectives on Science* 24, no. 5 (2016): 552–81.

67. Longino, *Science as Social Knowledge*. Kitcher, *Science in a Democratic Society*. Solomon, *Social Empiricism*. J. R. Brown, The community of science, in M. Carrier, D. Howard, and J. A. Kourany, eds., *The Challenge of the Social and the Pressure of Practice: Science and Values Revisited* (Pittsburgh, PA: University of Pittsburgh Press, 2008), 189–216.

68. We do not mean to imply that such accounts are useless or misguided; on the contrary, we believe they can offer important insights.

69. L. C. Hamilton, Education, politics and opinions about climate change evidence for interaction effects, *Climatic Change* 104, no. 2 (2011): 231–42. P. S. Hart and E. C. Nisbet, Boomerang effects in science communication: how motivated reasoning and identity cues amplify opinion polarization about climate mitigation policies, *Communication Research* 39, no. 6 (2012): 701–23. A. M. McCright, Political orientation moderates Americans' beliefs and concern about climate change, *Climatic Change* 104, no. 2 (2011): 243–53. A. M. McCright, R. E. Dunlap, and S. T. Marquart-Pyatt, Political ideology and views about climate change in the European Union, *Environmental Politics* 25, no. 2 (2016): 338–58. R. Gifford, The dragons of inaction: psychological barriers that limit climate change mitigation and adaptation, *American Psychologist* 66, no. 4 (2011): 290–302. Fielding and Hornsey, A social identity analysis. J. Sweetman and L. E. Whitmarsh, Climate justice: high-status ingroup social models increase pro-environmental action through making actions seem more moral, *Topics in Cognitive Science* 8, no. 1 (2016): 196–221. S. Lewandowsky, G. E. Gignac, and K. Oberauer, The role of conspiracist ideation and worldviews in predicting rejection of science, *PLOS One* 8, no. 10 (2013): e75637. Kahan, Ideology, motivated reasoning, and cognitive reflection. D. M. Kahan, H. Jenkins-Smith, and D. Braman, Cultural cognition of scientific consensus, *Journal of Risk Research* 14, no. 2 (2011): 147–74.

Chapter 2

1. J. R. Warren and B. J. Marshall, Unidentified curved bacilli on gastric epithelium in active chronic gastritis, *Lancet* 321, no. 8336 (1983): 1273–75.

2. B. J. Marshall and J. R. Warren, Unidentified curved bacilli in the stomach of patients with gastritis and peptic ulceration, *Lancet* 323, no. 8390 (1984): 1311–15.

3. M. A. Meyers, *Happy Accidents: Serendipity in Modern Medical Breakthroughs* (New York: Arcade, 2007).

4. National Institutes of Health, NIH consensus conference: *Helicobacter pylori* in peptic ulcer disease. NIH Consensus Development Panel on *Helicobacter pylori* in Peptic Ulcer Disease, *Journal of the American Medical Association* 272, no. 1 (1994): 65–69.

5. J. M. Lee, E. Deasy, and C. A. O'Morain, *Helicobacter pylori* eradication therapy: a discrepancy between current guidelines and clinical practice, *European Journal of Gastroenterology and Hepatology* 12, no. 4 (2000): 433–37.

6. E. Anderson, The epistemology of democracy, *Episteme: A Journal of Social Epistemology* 3, no. 1 (2006): 8–22. J. Beatty and A. Moore, Should we aim for consensus, *Episteme: A Journal of Social Epistemology* 7, no. 3 (2010): 198–214. P. Kitcher, *The Advancement of Science: Science without Legend, Objectivity without Illusions* (New York: Oxford University Press, 1993). H. E. Longino, *The Fate of Knowledge* (Princeton, NJ: Princeton University Press, 2002). H. E. Longino, *Science as Social Knowledge: Values and Objectivity in Scientific Inquiry* (Princeton, NJ: Princeton University Press, 1990). L. H. Nelson, *Who Knows: From Quine to a Feminist Empiricism* (Philadelphia, PA: Temple University Press, 1990). K. Popper, *Conjectures and Refutations: The Growth of Scientific Knowledge* (London: Routledge, 1963). P. Feyerbend, *Against Method: Outline of an Anarchistic Theory of Knowledge* (Atlantic Highlands, NJ: Humanities Press, 1975). M. Solomon, *Social Empiricism* (Cambridge, MA: MIT Press, 2001). P. Kitcher, *Science in a Democratic Society* (Amherst, NY: Prometheus, 2011).

7. Longino, *The Fate of Knowledge*. A. Richardson, Solomon's science without conscience, or, on the coherence of epistemic Newtonianism, *Perspective on Science* 16 (2008): 246–52.

8. J. S. Mill, *On Liberty* (Indianapolis, IN: Hackett, [1859] 1978).

9. Mill, *On Liberty*, 16.

10. T. S. Kuhn, *The Structure of Scientific Revolutions* (Chicago: University of Chicago Press, 1962). Longino, *Science as Social Knowledge*. N. R. Hanson, *Patterns of Discovery: An Inquiry into the Conceptual Foundations of Science* (Cambridge, UK: Cambridge University Press, 1958). P. M. M. Duhem, *The Aim and Structure of Physical Theory* (Princeton, NJ: Princeton University Press, [1906] 1954). W. V. O. Quine, Two dogmas of empiricism, in *From a Logical Point of View*, 2nd ed. (Cambridge, MA: Harvard University Press, 1951), 20–46.

11. Longino, *Science as Social Knowledge*. Longino, *The Fate of Knowledge*. Nelson, *Who Knows*. Popper, *Conjectures and Refutations*.

12. E. F. Keller, *A Feeling for the Organism: The Life and Work of Barbara McClintock* (San Francisco: W. H. Freeman, 1983).

13. B. McClintock, Mutable loci in maize, in *Yearbook 47* (Washington, DC: Carnegie Institution of Washington, 1947), 155–69. B. McClintock, The origin and behavior of mutable loci in maise, *Proceedings of the National Academy of Sciences USA* 36, no. 6 (1950): 344–55.

14. C. Biemont, A brief history of the status of transposable elements: from junk DNA to major players in evolution, *Genetics* 186, no. 4 (2010): 1085–93.

15. Keller, *A Feeling for the Organism.* S. Ravindran, Barbara McClintock and the discovery of jumping genes, *Proceedings of the National Academy of Sciences USA* 109, no. 50 (2012): 20198–99.

16. Kitcher, *Science in a Democratic Society.* Feyerbend, *Against Method.* Solomon, *Social Empiricism.* M. Solomon, Norms of epistemic diversity, *Episteme: A Journal of Social Epistemology* 3, no. 1 (2006): 23–36.

17. J. Strable and M. J. Scalon, Maize (*Zea mays*): a model organism for basic and applied research in plant biology, *Cold Spring Harbor Protocols* 4, no. 10 (2009).

18. Keller, *A Feeling for the Organism.*

19. Strable and Scalon, Maise.

20. S. L. Washburn and C. S. Lancaster, The evolution of hunting, in R. B. Lee and I. De Vore, eds., *Man the Hunter* (Chicago: Aldine, 1968), 293–303. W. S. Laughlin, Hunting: an integrating biobehavior system and its evolutionary importance, in R. B. Lee and I. DeVore, eds., *Man the Hunter* (Chicago: Aldine, 1968), 304–20. E. O. Wilson, *Sociobiology: The New Synthesis* (Cambridge, MA: Belknap Press of Harvard University Press, 1975).

21. A. Wylie, Doing social science as a feminist: the engendering of archaeology, in A. N. H. Creager, E. Lunbeck, and L. Schiebinger, eds., *Feminism in Twentieth Century Science, Technology, and Medicine* (Chicago: University of Chicago Press, 2001), 23–45.

22. A. Wylie and L. H. Nelson, Coming to terms with the values of science: insights from feminist science studies scholarship, in H. Kincaid, J. Dupre, and A. Wylie, eds., *Value-Free Science? Ideals and Illusions* (Oxford: Oxford University Press, 2007), 58–86. Wylie, Doing social science as a feminist. F. Dahlberg, *Woman the Gatherer* (New Haven, CT: Yale University Press, 1981). J. Gero, Genderlithics: women's roles in stone tool production, in J. Gero and M. W. Conkey, eds., *Engineering Archaeology* (Oxford: Oxford University Press, 1991), 163–92.

23. We examine these arguments in chapter 9 in more detail.

24. Longino, *Science as Social Knowledge.*

25. E. Anderson, Uses of value judgments in science: a general argument, with lessons from a case study of feminist research on divorce, *Hypatia* 19, no. 1 (2004): 1–24. I. de Melo-Martin and K. Intemann, Can ethical reasoning contribute to better epidemiology? A case study in research on racial health disparities, *European Journal of Epidemiology* 22, no. 4 (2007): 215–21. J. Dupré, Fact and value, in H. Kincaid, J. Dupre, and A. Wylie, eds., *Value-Free Science? Ideals and Illusions* (Oxford: Oxford University Press, 2007), 27–41. E. Martin, The egg and the sperm: how science has constructed a romance based on stereotypical male-female relationships, *Signs* 16, no. 3 (1991): 485–501. D. Ludwig, Ontological choices and the value-free ideal, *Erkenntnis* 81, no. 6 (2015): 1253–72.

26. K. Shrader-Frechette, *Risk and Rationality: Philosophical Foundations for Populist Reforms* (Berkeley: University of California Press, 1991). H. Douglas, Inductive risk and values in science, *Philosophy of Science* 67, no. 4 (2000): 559–79. H. Douglas, *Science, Policy, and the Value-Free Ideal* (Pittsburgh, PA: University of Pittsburgh Press, 2009).

27. W. Freeze and D. Schubert, Safety testing and regulation of genetically engineered foods, *Biotechnology and Genetic Engineering Reviews* 21 (2004): 299–324. E. Wickson and B. Wynne, Ethics of science for policy in the environmental governance of biotechnology: MON819 maize in Europe, *Ethics, Policy and Environment* 15, no. 3 (2012): 321–40.

28. Wickson and Wynne, Ethics of science for policy.

29. T. Bohn et al., Reduced fitness of *Daphnia magna* fed a Bt-transgenic maize variety, *Archives of Environmental Contamination and Toxicology* 55, no. 4 (2008): 584–92.

30. P. E. Graves, *Environmental Economics: A Critique of Benefit-Cost Analysis* (Lanham, MD: Roman & Littlefield, 2007).

31. S. Schneider, Global warming: neglecting the complexities, *Scientific American* 286, no. 1 (2002): 62–65. A. Agarwal, A southern perspective on curbing global climate change, in S. H. Schneider, A. Rosencranz, and J. O. Niles, eds., *Climate Change Policy: A Survey* (Washington, DC: Island Press, 2002), 375–91.

32. R. M. Adams et al., Effects of global climate change on agriculture: an interpretative review, *Climate Research* 11, no. 1 (1998): 19–30.

33. Adams et al., Effects of global climate change. M. Parry, C. Rosenzweig, and M. Livermore, Climate change, and risk global food supply of hunger, *Philosophical Transactions of the Royal Society–Biological Sciences* 360, no. 1463 (2005): 2125–38.

34. Agarwal, A southern perspective.

35. Agarwal, A southern perspective. Q. Schiermeier, The real holes in climate science, *Nature* 463, no. 7279 (2010): 284–87. T. Wheeler and J. von Braun, Climate change impacts on global food security, *Science* 341, no. 6145 (2013): 508–13.

36. Wakefield et al., Ileal-lymphoid-nodular hyperplasia. D. A. Geier and M. R. Geier, A meta-analysis epidemiological assessment of neurodevelopmental disorders following vaccines administered from 1994 through 2000 in the United States, *Neuroendocrinology Letters* 27, no. 4 (2006): 401–13.

37. For example, B. S. Gadad et al., Administration of thimerosal-containing vaccines to infant rhesus macaques does not result in autism-like behavior or neuropathology, *Proceedings of the National Academy of Sciences USA* 112, no. 49 (2015): E6827. A. Jain et al., Autism occurrence by MMR vaccine status among U.S. children with older siblings with and without autism, *Journal of the American Medical Association* 31, no. 15 (2015): 1534–40. C. Mellis, Measles-mumps-rubella vaccine does not cause autism, *Journal of Paediatrics and Child Health* 51, no. 8 (2015): 838. Y. Uno et al., Early exposure to the combined measles-mumps-rubella vaccine and thimerosal-containing vaccines and risk of autism spectrum disorder, *Vaccine* 33, no. 21 (2015): 2511–16. L. E. Taylor et al., Vaccines are not associated with autism: an evidence-based meta-analysis of case-control and cohort studies, *Vaccine* 32, no. 29 (2014): 3623–29. C. S. Price et al., Prenatal and infant exposure to thimerosal from vaccines and immunoglobulins and risk of autism, *Pediatrics* 126, no. 4 (2010): 656–64. A. Cox, A. Taliercio, and C. Zacharyczuk, Additional CDC data spell vaccine-autism link, *Pediatric Annals* 42, no. 5 (2013): 178.

38. Anderson, Epistemology of democracy. Beatty and Moore, Should we aim for consensus. Kitcher, *Advancement of Science.*

39. P. J. Michaels, How to manufacture a climate consensus, *Wall Street Journal,* December 17, 2009. J. P. van der Sluijs, R. van Est, and M. Riphagen, Beyond consensus: reflections from a democratic perspective on the interaction between climate politics and science, *Current Opinion in Environmental Sustainability* 2, nos. 5-6 (2010): 409–15. J. A. Curry and P. J. Webster, Climate science and the uncertainty monster, *Bulletin of the American Meteorological Society* 92, no. 12 (2011): 1667–82.

40. Longino, *The Fate of Knowledge.* Longino, *Science as Social Knowledge.*

41. J. Bohannon, Who's afraid of peer review? *Science* 342, no. 6154 (2013): 60–65. C. J. Haug, Peer-review fraud: hacking the scientific publication process, *New England Journal*

of Medicine 373, no. 25 (2015): 2393–95. C. J. Lee et al., Bias I peer review, *Journal of the American Society for Information Science and Technology* 64, no. 1 (2013): 2–17. C. J. Lee, Commensuration bias in peer review, *Philosophy of Science* 82, no. 5 (2015): 1272–83.

42. K. Intemann, Why diversity matters: understanding and applying the diversity component of the National Science Foundation's broader impacts criterion, *Social Epistemology* 23, nos. 3–4 (2009): 249–66. B. Miller, When is consensus knowledge-based? Distinguishing shared knowledge from mere agreement, *Synthese* 190, no. 7 (2013): 1293–316.

43. Longino, *Science as Social Knowledge*.

44. A. Wylie, Why standpoint matters, in R. Figueroa and S. Harding, eds., *Science and Other Cultures: Issues in Philosophies of Science and Technology* (New York: Routledge, 2003), 26–48. Intemann, Why diversity matters.

45. Solomon, Norms of epistemic diversity.

46. K. Intemann and I. de Melo-Martin, Social values and scientific evidence: the case of the HPV vaccines, *Biology and Philosophy* 25, no. 2 (2010): 203–13. I. de Melo-Martin and K. Intemann, Feminist resources for biomedical research: lessons from the HPV vaccines, *Hypatia* 26, no. 1 (2011): 79–101.

47. Longino, *Science as Social Knowledge*. Longino, *The Fate of Knowledge*.

48. I. Loudon, *The Tragedy of Childbed Fever* (Oxford and New York: Oxford University Press, 2000). S. B. Nuland, *The Doctors' Plague: Germs, Childbed Fever, and the Strange Story of Ignac Semmelweis* (New York: W.W. Norton, 2003).

49. Loudon, *The Tragedy of Childbed Fever*. Nuland, *The Doctors' Plague*.

50. W. Goodwin, Revolution and progress in medicine, *Theoretical Medicine and Bioethics* 36, no. 1 (2015): 25–39. D. Tulodziecki, Shattering the myth of Semmelweis, *Philosophy of Science* 80, no. 5 (2013): 1065–75.

51. Longino, *The Fate of Knowledge*. Longino, *Science as Social Knowledge*.

52. V. C. Rubin et al., Rotation velocities of 16 SA galaxies and a comparison of Sa, Sb, and Sc rotation properties, *Astrophysical Journal* 289 (1985): 81–98, 101–104. V. C. Rubin, W. Ford, and J. Kent, Rotation of the Andromeda nebula from a spectroscopic survey of emission regions, *Astrophysical Journal* 159 (1970): 379–403.

53. V. C. Rubin, Differential rotation of the inner metagalaxy, *Astronomical Journal* 56, no. 2 (1951): 47–48. V. C. Rubin, Fluctuations in the space distribution of the galaxies, *Proceedings of the National Academy of Sciences USA* 40, no. 7 (1954): 541–49.

54. Rubin, Differential rotation. Rubin, Fluctuations in the space distribution.

55. V. C. Rubin, Am interesting voyage, *Annual Review of Astronomy and Astrophysics* 49 (2011): 1–28.

56. Rubin, Fluctuations in the space distribution.

57. S. Crasnow, Is standpoint theory a resource for feminist epistemology? An introduction, *Hypatia* 24, no. 4 (2009): 189–92. S. G. Harding, A socially relevant philosophy of science? Resources from standpoint theory's controversiality, *Hypatia* 19, no. 1 (2004): 25–47. Wylie, Why standpoint matters. H. Longino, Gender, politics, and the theoretical virtues, *Synthese* 104, no. 3 (1995): 383–97.

58. Longino, *The Fate of Knowledge*.

59. Fricker, *Epistemic Injustice*.

60. Harding, *Sciences from Below*. De Mello-Martin and Intemann, Feminist resources. P. Nadasdy, *Hunters and Bureaucrats: Power, Knowledge, and Aboriginal-State Relations in the Southwest Yukon* (Vancouver: University of British Columbia Press, 2003). F. Mazzocchi, Western science and traditional knowledge: despite their variations, different forms of

knowledge can learn from each other, *EMBO Reports* 7, no. 5 (2006): 463–66. M. Iaccarino, Science and culture: Western science could learn a thing or two from the way science is done in other cultures, *EMBO Reports* 4, no. 3 (2003): 220–23.

61. S. Epstein, *Impure Science: AIDS, Activism, and the Politics of Knowledge* (Berkeley: University of California Press, 1996).

62. Epstein, *Impure Science.*

63. Epstein, *Impure Science.*

64. Harding, *Sciences from Below.*

65. N. A. McHugh, *The Limits of Knowledge: Generating Pragmatist Feminist Cases for Situated Knowing* (Albany: State University of New York Press, 2015). M. Gough, *Dioxin, Agent Orange: The Facts* (New York: Plenum Press, 1986).

66. J. M. Friedman, Does Agent Orange cause birth defects? *Teratology* 29, no. 2 (1984): 193–221. J. D. Erickson et al., Vietnam veterans' risks for fathering babies with birth defects, *Journal of the American Medical Association* 252, no. 7 (1984): 903–12. J. P. Whitlock, The regulation of gene-expression by 2,3,7,8-tetrachlorodibenzo-para-dioxin, *Pharmacological Reviews* 39, no. 2 (1987): 147–61. W. H. Wolfe et al., Paternal serum dioxin and reproductive outcomes among veterans of Operation Ranch Hand, *Epidemiology* 6, no. 1 (1995): 17–22. A. D. Ngo et al., Association between Agent Orange and birth defects: systematic review and meta-analysis, *International Journal of Epidemiology* 35, no. 5 (2006): 1220–30. Gough, *Dioxin, Agent Orange.*

67. K. M. Stevens, Agent Orange toxicity: a quantitative perspective, *Human Toxicology* 1, no. 1 (1981): 31–39. Whitlock, The regulation of gene-expression. Wolfe et al., Paternal serum dioxin. Friedman, Does Agent Orange cause birth defects? Erickson et al., Vietnam veterans' risks.

68. McHugh, *The Limits of Knowledge.*

69. Ngo et al., Association between Agent Orange and birth defects. McHugh, *The Limits of Knowledge.*

70. McHugh, *The Limits of Knowledge.* A. Schecter et al., Recent dioxin contamination from Agent Orange in residents of a southern Vietnam city, *Journal of Occupational and Environmental Medicine* 43, no. 5 (2001): 435–43.

71. T. N. Lee and A. Johansson, Impact of chemical warfare with Agent Orange on women's reproductive lives in Vietnam: a pilot study, *Reproductive Health Matters* 9, no. 10 (2001): 156–64.

72. A. Schecter et al., A follow-up: high level of dioxin contamination in Vietnamese from Agent Orange, three decades after the end of spraying, *Journal of Occupational and Environmental Medicine* 44, no. 3 (2002): 218–20. Ngo et al., Association between Agent Orange and birth defects.

73. McHugh, *The Limits of Knowledge.*

74. McHugh, *The Limits of Knowledge.*

75. J. A. Kourany, *Philosophy of Science after Feminism* (Oxford and New York: Oxford University Press, 2010). S. G. Harding, *Objectivity and Diversity: Another Logic of Scientific Research* (Chicago: University of Chicago Press, 2015). L. L. Schiebinger, *Plants and Empire: Colonial Bioprospecting in the Atlantic World* (Cambridge, MA: Harvard University Press, 2004). L. Code, *Ecological Thinking: The Politics of Epistemic Location* (Oxford and New York: Oxford University Press, 2006). McHugh, *The Limits of Knowledge.* A. Fausto-Sterling, *Myths of Gender: Biological Theories about Women and Men* (New York: Basic Books, 1992). Wylie, Why standpoint matters.

76. S. B. Hrdy, Empathy, polyandry and the myth of the coy female, in R. Bleier, ed., *Feminist Approaches to Science* (New York: Pergamon, 1986), 119–46.

77. Wilson, *Sociobiology*. R. L. Trivers, Parental investment and sexual selection, in B. Campbell, ed., *Sexual Selection and the Descent of Man, 1871–1971* (Chicago: Aldine, 1972), 136–79.

78. Hrdy, Empathy, polyandry.

79. Hrdy, Empathy, polyandry.

Chapter 3

1. D. Michaels, *Doubt Is Their Product: How Industry's Assault on Science Threatens Your Health* (Oxford and New York: Oxford University Press, 2008). N. Oreskes and E. M. Conway, *Merchants of Doubt: How a Handful of Scientists Obscured the Truth on Issues from Tobacco Smoke to Global Warming* (New York: Bloomsbury, 2010). R. Proctor, *Golden Holocaust: Origins of the Cigarette Catastrophe and the Case for Abolition* (Berkeley: University of California Press, 2011).

2. Michaels, *Doubt*, 11.

3. Oreskes and Conway, *Merchants of Doubt*.

4. Oreskes and Conway, *Merchants of Doubt*. Michaels, *Doubt*. T. O. McGarity and W. Wagner, *Bending Science: How Special Interests Corrupt Public Health Research* (Cambridge, MA: Harvard University Press, 2008).

5. K. Shrader-Frechette, Conceptual analysis and special-interest science: toxicology and the case of Edward Calabrese, *Synthese* 177, no. 3 (2010): 449–69. P. Mushak and K. C. Elliott, Structured development and promotion of a research field: hormesis in biology, toxicology, and environmental regulatory science, *Kennedy Institute of Ethics Journal* 225, no. 4 (2015): 335–67.

6. R. Brulle, Institutionalizing delay: foundation funding and the creation of U.S. climate change counter-movement organizations, *Climatic Change* 122, no. 4 (2014): 681–94.

7. I. Sample, Scientists offered cash to dispute climate study, *The Guardian*, February 2, 2007.

8. Proctor, *Golden Holocaust*. Michaels, *Doubt*. McGarity and Wagner, *Bending Science*.

9. We discuss the value-free ideal in more detail in chapter 9.

10. E. McMullin, Values in science, in P. Asquith and T. Nickels, eds., *PSA 1982: The Proceedings of the 1982 Biennial Meeting of the Philosophy of Science Association*, 1: 3–28 (East Lansing, MI: Philosophy of Science Association, 1983). L. Laudan, The epistemic, the cognitive, and the social, in P. K. Machamer and G. Wolters, eds., *Science, Values, and Objectivity* (Pittsburgh, PA: University of Pittsburgh Press, 2004), 14–23. D. Steel, Epistemic values and the argument from inductive risk, *Philosophy of Science* 77, no. 1 (2010): 14–34.

11. G. Betz, In defense of the value free ideal, *European Journal of the Philosophy of Science* 3, no. 2 (2013): 207–20. M. Dorato, Epistemic and nonepistemic values in science, in P. Machamer and G. Wolters, eds., *Science, Values, and Objectivity* (Pittsburgh, PA: University of Pittsburgh Press, 2004), 52–77.

12. H. Douglas, *Science, Policy, and the Value-Free Ideal* (Pittsburgh, PA: University of Pittsburgh Press, 2009). S. Haack, *Manifesto of a Passionate Moderate: Unfashionable Essays* (Chicago: University of Chicago Press, 1998). E. Anderson, Uses of value judgments in

science: a general argument, with lessons from a case study of feminist research on divorce, *Hypatia* 19, no. 1 (2004): 1–24.

13. P. Kitcher, *Science, Truth, and Democracy* (Oxford and New York: Oxford University Press, 2001). J. A. Kourany, *Philosophy of Science after Feminism* (Oxford and New York: Oxford University Press, 2010). B. Miller, Science, values, and pragmatic encroachment on knowledge, *European Journal for Philosophy of Science* 4, no. 2 (2014): 253–70. J. Dupré, Fact and value, in H. Kincaid, J. Dupre, and A. Wylie, eds., *Value-Free Science? Ideals and Illusions* (Oxford: Oxford University Press, 2007), 27–41.

14. P. Kitcher, *The Advancement of Science: Science without Legend, Objectivity without Illusions* (New York: Oxford University Press, 1993). M. Carrier and A. Nordmann, *Science in the Context of Application*, in Boston Studies in the Philosophy of Science series (Dordrecht and London: Springer, 2011), 274. H. E. Longino, *Science as Social Knowledge: Values and Objectivity in Scientific Inquiry* (Princeton, NJ: Princeton University Press, 1990).

15. J. Reiss, In favour of a Millian proposal to reform biomedical research, *Synthese* 177, no. 3 (2010): 427–47.

16. Kourany, *Philosophy of Science after Feminism*.

17. Kitcher, *Advancement of Science*.

18. Proctor, *Golden Holocaust*, 430.

19. D. Harker, *Creating Scientific Controversies: Uncertainty and Bias in Science and Society* (Cambridge, UK: Cambridge University Press, 2015).

20. Harker, *Creating Scientific Controversies*. N. Oreskes, The scientific consensus on climate change: how do we know we're not wrong?, in J. DiMento and P. Doughman, eds., *Climate Change: What It Means for Us, Our Children, and Our Grandchildren* (Cambridge, MA: MIT Press, 2007), 65–99. M. Solomon, *Social Empiricism* (Cambridge, MA: MIT Press, 2001).

21. We are not suggesting that all these authors are explicitly concerned with proposing indicators that can help us identify bad faith motives—and thus problematic dissent. In *Creating Scientific Controversies*, David Harker does explicitly propose the features we discuss here as indictors that the dissent is offered in bad faith (although he does not refer to bad-faith dissent but to created controversies). Rather, when discussing dissent they take to be problematic, these authors point out to features of such dissent that we believe one could use as suggesting that the dissent is offered in bad faith. Our purpose here is to assess whether such is the case.

22. R. Horton, Vioxx, the implosion of Merck, and aftershocks at the FDA, *Lancet* 364, no. 9450 (2004): 1995–96. K. Thomas, In documents on pain drug, signs of doubt and deception, *New York Times*, June 24, 2012.

23. N. Nkansah et al., Randomized trials assessing calcium supplementation in healthy children: relationship between industry sponsorship and study outcomes, *Public Health Nutrition* 12, no. 10 (2009): 1931–37. S. Sismondo, Pharmaceutical company funding and its consequences: a qualitative systematic review, *Contemporary Clinical Trials* 29, no. 2 (2008): 109–13. S. N. Khan et al., The roles of funding source, clinical trial outcome, and quality of reporting in orthopedic surgery literature, *American Journal of Orthopedics* 37, no. 12 (2008): E205–12. F. T. Bourgeois, S. Murthy, and K. D. Mandl, Outcome reporting among drug trials registered in ClinicalTrials.gov, *Annals of Internal Medicine* 153, no. 3 (2010): 158–66. A. Lundh et al., Industry sponsorship and research outcome, *Cochrane Database of Systemic Reviews* 12 (2012): MR000033.

24. A. W. Jørgensen, J. Hilden, and P.C. Gøtzsche, Cochrane reviews compared with industry supported meta-analyses and other meta-analyses of the same drugs: systematic review, *British Medical Journal* 333, no. 7572 (2006): 782. A. C. Tricco et al., Non-Cochrane vs. Cochrane reviews were twice as likely to have positive conclusion statements: cross-sectional study, *Journal of Clinical Epidemiology* 62, no. 4 (2009): 380–86.e381.

25. X. Sun et al., The influence of study characteristics on reporting of subgroup analyses in randomised controlled trials: systematic review, *British Medical Journal* 342 (2011): d1569.

26. S. Krimsky, *Science in the Private Interest: Has the Lure of Profits Corrupted Biomedical Research?* (Lanham, MD: Rowman & Littlefield, 2003). H. Brody, *Hooked: Ethics, the Medical Profession, and the Pharmaceutical Industry* (Lanham, MD: Rowman & Littlefield, 2007). I. de Melo-Martín and K. Intemann, How do disclosure policies fail? Let us count the ways, *Federation of American Societies for Experimental Biology Journal* 23, no. 6 (2009): 1638–42.

27. I. de Melo-Martín and K. Intemann, How do disclosure policies fail? K. C. Elliott, Scientific judgment and the limits of conflict-of-interest policies, *Accountability in Research* 15, no. 1 (2008): 1–29. M. Carrier, Science in the grip of the economy, in M. Carrier, D. Howard, and J. A. Kourany, eds., *The Challenge of the Social and the Pressure of Practice: Science and Values Revisited* (Pittsburgh, PA: University of Pittsburgh Press, 2008), 217–34.

28. B. Zycher, J. DiMasi, and C. Milne, Private sector contributions to pharmaceutical science: thirty-five summary case histories, *American Journal of Therapeutics* 17, no. 1 (2010): 101–20.

29. M. J. Goldenberg, Public misunderstanding of science? Reframing the problem of vaccine hesitancy, *Perspectives on Science* 24, no. 5 (2016): 552–81.

30. E. Lipton, Food industry enlisted academics in G.M.O. lobbying war, emails show, *New York Times*, September 5, 2015. https://www.nytimes.com/2015/09/06/us/food-industry-enlisted-academics-in-gmo-lobbying-war-emails-show.html.

31. P. Bagla and R. K. Pachauri, Climate science leader Rajendra Pachauri confronts the critics, *Science* 327, no. 5965 (2010): 510–11. R. A. Pielke, Major change is needed if the IPCC hopes to survive, *Yale Environment 360*, February 25, 2010. http://e360.yale.edu/feature/major_change_is_needed_if_the_ipcc_hopes_to_survive/2244/. R. Mendick, Taxpayers' millions paid to Indian institute run by UN climate chief, *The Telegraph,* January 16, 2010. http://www.telegraph.co.uk/news/earth/environment/climatechange/7005963/Taxpayers-millions-paid-to-Indian-institute-run-by-UN-climate-chief.html.

32. In chapter 8, we consider the negative effects that financial conflicts of interests have on warranted public trust in the scientific community.

33. Harker, *Creating Scientific Controversies*. Oreskes, The scientific consensus on climate change.

34. Harker, *Creating Scientific Controversies*.

35. J. M. Inhofe, *The Greatest Hoax: How the Global Warming Conspiracy Threatens Your Future* (Washington, DC: WND Books, 2012). S. McIntyre, *Climate Audit.* (2016). https://climateaudit.org/. Friends of Science (FoS), President's message: FoS is dedicated to providing the public with insight into climate science, *FoS Membership Quarterly Newsletter* 50 (2016). https://friendsofscience.org/assets/documents/2016_June_Newsletter.pdf.

36. Oreskes, The scientific consensus on climate change, 75.

37. National Vaccine Information Center (NVIC). (2016). www.nvic.org/. A. J. Wakefield, director, *Vaxxed: From Cover-Up to Catastrophe* (film, 2016; produced by Del Bigtree, distributed by Cinema Libre Studio). L. K. Habakus an.d M. Holland, *Vaccine Epidemic: How Corporate Greed, Biased Science, and Coercive Government Threaten Our Human Rights, Our Health, and Our Children* (New York: Skyhorse, 2011).

38. Harker, *Creating Scientific Controversies.*

39. C. J. Lee et al., Bias in peer review, *Journal of the American Society for Information Science and Technology* 64, no. 1 (2013): 2–17. T. Luukkonen, Conservatism and risk-taking in peer review: emerging ERC practices, *Research Evaluation* 21, no. 1 (2012): 48–60.

40. Luukkonen, Conservatism and risk-taking in peer review. K. I. Resch, E. Ernst, and J. Garrow, A randomized controlled study of reviewer bias against an unconventional therapy, *Journal of the Royal Society of Medicine* 93, no. 4 (2000): 164–67.

41. Lee et al., Bias in peer review. M. Jelicic and H. Merckelbach, Peer review: let's imitate the lawyers! *Cortex* 38, no. 3 (2002): 406–407. R. Nickerson, Confirmation bias: a ubiquitous phenomenon in many guises, *Review of General Psychology* 2, no. 2 (1998): 175–220.

42. F. Pearce, Climate change emails between scientists real flaws in peer review, *The Guardian*, February 2, 2010. https://www.theguardian.com/environment/2010/feb/02/hacked-climate-emails-flaws-peer-review. P. J. Michaels, How to manufacture a climate consensus, *Wall Street Journal*, December 17, 2009. www.wsj.com/articles/SB10001424052748704398304574598230426037244.

43. M. J. Behe, *The Edge of Evolution: The Search for the Limits of Darwinism* (New York: Free Press, 2007). W. A. Dembski and J. Wells, *The Design of Life: Discovering Signs of Intelligence in Biological Systems* (Dallas, TX: Foundation for Thought and Ethics, 2008). J. Witt, The intelligent approach: teach the strengths and weakness of evolution (2005), Discovery Institute website, posted June 17, 2013. www.discovery.org/a/2743.

44. P. Nadasdy, *Hunters and Bureaucrats: Power, Knowledge, and Aboriginal-State Relations in the Southwest Yukon* (Vancouver: University of British Columbia Press, 2003).

45. Nadasdy, *Hunters and Bureaucrats.* J. Kochan, Objective styles in northern field science, *Studies in History and Philosophy of Science* 52 (2015): 1–12.

46. Nadasdy, *Hunters and Bureaucrats.*

47. Kochan, Objective styles in northern field science. J. H. Schmidt et al., Using distance sampling and hierarchical models to improve estimates of Dall's sheep abundance, *Journal of Wildlife Management* 76, no. 2 (2012): 317–27. K. L. Monteith et al., Effects of harvest, culture, and climate on trends in size of horn-like structures in trophy ungulates, *Wildlife Monographs* 183 (2013): 1–28.

48. P. Kl_progge and J. P. van der Sluijs, The inclusion of stakeholder knowledge and perspectives in integrated assessment of climate change, *Climatic Change* 75, no. 3 (2006): 359–89. J. Douglas, *Science, Policy, and the Value-Free Ideal* (Pittsburgh, PA: University of Pittsburgh Press, 2009). K. Shrader-Frechette, *Taking Action, Saving Lives: Our Duties to Protect Environmental and Public Health* (Oxford and New York: Oxford University Press, 2007). K. Intemann, Distinguishing between legitimate and illegitimate values in climate modeling, *European Journal of Philosophy of Science* 5, no. 2 (2015): 217–32.

49. J. Romeis et al., When bad science makes good headlines: Bt maize and regulatory bans, *Nature Biotechnology* 31, no. 5 (2013): 386–87. A. Nicolia et al., An overview of the last 10 years of genetically engineered crop safety research, *Critical Reviews in Biotechnology* 34, no. 1 (2014): 77–88.

50. M. Cuhra, Review of GMO safety assessment studies: glyphosate residues in Roundup Ready crops is an ignored issue, *Environmental Sciences Europe* 27 (2015): 20. J. A. Magana-Gomez and A. M. C. de la Barca, Risk assessment of genetically modified crops for nutrition and health, *Nutrition Reviews* 67, no. 1 (2009): 1–16. W. Zhang and F. Shi, Do genetically modified crops affect animal reproduction? A review of the ongoing debate, *Animal* 5, no. 7 (2011): 1048–59. D. F. Bushey et al., Characteristics and safety assessment of intractable proteins in genetically modified crops, *Regulatory Toxicology and Pharmacology* 69, no. 2 (2014): 154–70.

51. Harker, *Creating Scientific Controversies*. Solomon, *Social Empiricism*. Oreskes, The scientific consensus on climate change.

52. Oreskes, The scientific consensus on climate change. Harker, *Creating Scientific Controversies*.

53. Harker, *Creating Scientific Controversies*, 163.

54. K. Shrader-Frechette, Hydrogeology and framing questions having policy consequences, *Philosophy of Science* 64, no. 4 (1997): S149–60. N. Oreskes, K. Shrader-Frechette, and K. Belitz, Verification, validation, and confirmation of numerical models in the earth sciences, *Science* 263, no. 5147 (1994): 641–46.

55. J. A. Curry and P. J. Webster, Climate science and the uncertainty monster, *Bulletin of the American Meteorological Society* 92, no. 12 (2011): 1667–82. E. Hawkins et al., Irreducible uncertainty in near-term climate projections, *Climate Dynamics* 46, no. 11 (2015): 3807–19.

56. N. Pidgeon, Climate change risk perception and communication: addressing a critical moment? *Risk Analysis* 32, no. 6 (2012): 951–56. S. Dessai and M. Hulme, Does climate adaptation policy need probabilities? *Climate Policy* 4, no. 2 (2004): 107–28.

57. R. Dunlap and A. McCright, Climate change denial: sources, actors and strategies, in C. Lever-Tracy, ed., *Routledge Handbook of Climate Change and Society* (London: Routledge, 2010), 240–59. A. J. Hoffman, Talking past each other? Cultural framing of skeptical and convinced logics in the climate change debate, *Organization and Environment* 24, no. 1 (2011): 3–33. B. B. Johnson, Climate change communication: a provocative inquiry into motives, meanings, and means, *Risk Analysis* 32(6) (2012): 973–91. S. Lewandowsky et al., The role of conspiracist ideation and worldviews in predicting rejection of science, *PLOS One* 8, no. 10 (2013): e75637. A. Gelfert, Climate scepticism, epistemic dissonance, and the ethics of uncertainty, *Philosophy and Public Issues* 3, no. 1 (2013): 168–208.

58. Hoffman, Talking past, 5.

59. Harker, *Creating Scientific Controversies*; L. Ceccarelli, Manufactured scientific controversy: science, rhetoric, and public debate, *Rhetoric and Public Affairs* 14, no. 2 (2011): 195–228.

60. A. J. Wakefield et al. Ileal-lymphoid-nodular hyperplasia, non-specific colitis, and pervasive developmental disorder in children [retracted], *Lancet* 351, no. 9103 (1998): 637–41. B. Deer, Secrets of the MMR scare: how the case against the MMR vaccine was fixed, *British Medical Journal* 342 (2011): c5347. B. Deer, Secrets of the MMR scare: how the vaccine crisis was meant to make money, *British Medical Journal* 342 (2011): c5258.

61. B. S. Gadad et al., Administration of thimerosal-containing vaccines to infant rhesus macaques does not result in autism-like behavior or neuropathology, *Proceedings*

of the National Academy of Sciences USA 112, no. 49 (2015): E6827. L. E. Taylor, A. L. Swerdfeger, and G. D. Eslick, Vaccines are not associated with autism: an evidence-based meta-analysis of case-control and cohort studies, *Vaccine* 32, no. 29 (2014): 3623–29.

Some might counter that vaccine dissent has been epistemically detrimental to the field of autism research because the focus on vaccines has diverted resources away from efforts to understand the real causes of autism (G. P. Oakley and R. B. Johnston, Balancing benefits and harms in public health prevention programmes mandated by governments, *British Medical Journal* 329, no. 7456 [2004]: 41–43). We do not deny this possibility. Our point is not to claim that bad-faith dissent is harmless, only that the presence of bad-faith motives need not be inconsistent with knowledge production; and thus, as a criterion for NID it is unreliable.

62. Harker, *Creating Scientific Controversies*, 2015.

Chapter 4

1. J. Delborne, Transgenes and transgressions: scientific dissent as heterogeneous practice, *Social Studies of Science* 38, no. 4 (2008): 509–41.

2. T. S. Kuhn, *The Essential Tension: Selected Studies in Scientific Tradition and Change* (Chicago: University of Chicago Press, 1977). H. E. Longino, *Science as Social Knowledge: Values and Objectivity in Scientific Inquiry* (Princeton, NJ: Princeton University Press, 1990). H. E. Longino, *The Fate of Knowledge* (Princeton, NJ: Princeton University Press, 2002). M. Solomon, *Social Empiricism* (Cambridge, MA: MIT Press, 2001). P. Kitcher, *Science in a Democratic Society* (Amherst, NY: Prometheus, 2011). K. Borgerson, Amending and defending critical contextual empiricism, *European Journal for Philosophy of Science* 1, no. 3 (2011): 435–49.

3. Longino, *Science as Social Knowledge*. Longino, *The Fate of Knowledge*. Kitcher, *Science in a Democratic Society*.

4. Longino, *The Fate of Knowledge*. D. Harker, *Creating Scientific Controversies: Uncertainty and Bias in Science and Society* (Cambridge, UK: Cambridge University Press, 2015). Longino (in *Science as Social Knowledge*) initially proposed these three criteria not as criteria of NID but, together with open avenues for criticism, as criteria for the objectivity of scientific communities. The goal of inquiry in her account, however, is to produce transformative criticism, and she makes clear in later work that criticism or dissent that does not meet these criteria would fail to be transformative in the appropriate way and may even be used to limit the sorts of obligations that we have to dissenters (Longino, *The Fate of Knowledge*).

5. These norms apply not only to dissenters but also to proponents of a consensus view. Because our concern here is with dissent, we refer mainly to dissenters.

6. Longino, *Science as Social Knowledge*. Longino, *The Fate of Knowledge*. Kitcher, *Science in a Democratic Society*.

7. Longino, *The Fate of Knowledge*. Harker, *Creating Scientific Controversies*. J. Cook et al., Consensus on consensus: a synthesis of consensus estimates on human-caused global warming, *Environmental Research Letters* 11, no. 4 (2016): 048002.

8. Longino, *Science as Social Knowledge*. Longino, *The Fate of Knowledge*. L. H. Nelson, *Who Knows: From Quine to a Feminist Empiricism* (Philadelphia, PA: Temple University Press, 1990).

9. Kitcher, *Science in a Democratic Society*. Longino, *The Fate of Knowledge*. M. Solomon, *Norms of Dissent*. Centre for the Philosophy of Natural and Social Science Contingency and Dissent in Science. Technical report 09/08, 2–21, 2008. www.lse. ac.uk/CPNSS/projects/CoreResearchProjects/ContingencyDissentInScience/DP/ SolomonNormsOfDissent0908Online.pdf.

10. Solomon, *Norms of Dissent*. Kitcher, *Science in a Democratic Society*. J. Biddle and A. Leuschner, Climate skepticism and the manufacture of doubt: Can dissent in science be epistemically detrimental? *European Journal for Philosophy of Science* 5, no. 3 (2015): 261–78.

11. N. Oreskes. The scientific consensus on climate change: how do we know we're not wrong? in J. DiMento and P. Doughman, eds., *Climate Change: What It Means for Us, Our Children, and Our Grandchildren* (Cambridge, MA: MIT Press, 2007), 65–99. N. Oreskes and E. M. Conway, *Merchants of Doubt: How a Handful of Scientists Obscured the Truth on Issues from Tobacco Smoke to Global Warming* (New York: Bloomsbury, 2010). Harker, *Creating Scientific Controversies*. Cook et al., Consensus on consensus.

12. Some caveats are in order here. First, we are not suggesting that there are not important differences between these cases of dissent—differences that might affect how scientific communities should approach them. For example, one might reasonably argue that there is more uncertainty regarding the cases of climate change science and GMPs than there is in the case of evolutionary theory and that such uncertainties are relevant when determining whether we have a case of NID. Such differences indeed exist and might well be relevant. However, we use these cases here as equivalent because many take these as paradigmatic cases of NID. Second, we are not claiming that these cases do in fact constitute cases of NID (or that it is uncontroversial that these examples should be categorized in this way). Although many agree that intelligent-design views fail to promote scientific progress—and thus amount to NID—significant disagreement exists about whether dissent in the case of GMP safety is indeed normatively inappropriate. As we said in chapter 1, our concern is not to demonstrate that these are or are not instances of NID. Rather, we use these cases—again, cases that many have taken to involve NID—to assess how they fare regarding the various criteria that have been proposed to identify problematic dissent. The fact that there is reasonable disagreement about which are clear cases of NID supports our overarching argument in this book that justifying and applying criteria for NID are difficult tasks.

13. Longino, *Science as Social Knowledge*. Longino, *The Fate of Knowledge*. Nelson, *Who Knows*. Kitcher, *Science in a Democratic Society*.

14. Kitcher, *Science in a Democratic Society*. Longino, *Science as Social Knowledge*. Longino, *The Fate of Knowledge*.

15. Longino, *The Fate of Knowledge*. Kitcher, *Science in a Democratic Society*.

16. S. J. Gould, *Ontogeny and Phylogeny* (Cambridge, MA: Belknap Press of Harvard University Press, 1977). M. Ruse, *The Evolution-Creation Struggle* (Cambridge, MA: Harvard University Press, 2005). W. A. Dembski and M. Ruse, *Debating Design: From Darwin to DNA* (New York: Cambridge University Press, 2004). P. Kitcher, *Living with Darwin: Evolution, Design, and the Future of Faith* (Oxford and New York: Oxford University Press, 2007). S. Sarkar, *Doubting Darwin? Creationist Designs on Evolution* (Malden, MA, and Oxford: Blackwell, 2007). R. Dawkins, *The Greatest Show on Earth: The Evidence for Evolution* (New York: Free Press, 2009). J. A. Coyne, *Why Evolution Is True*

(New York: Viking, 2009). J. Brockman, *Intelligent Thought: Science versus the Intelligent Design Movement* (New York: Vintage Books, 2006).

17. Kitcher, *Science in a Democratic Society*. Ruse, *The Evolution-Creation Struggle*.

18. D. T. Gish, *Evolution: The Challenge of the Fossil Record* (El Cajon, CA: Creation-Life, 1985). G. Sewell, *In the Beginning: And Other Essays on Intelligent Design* (Seattle, WA: Discovery Institute Press, 2010). M. J. Behe, *Darwin's Black Box: The Biochemical Challenge to Evolution* (New York: Free Press, 1996). W. A. Dembski, *Intelligent Design: The Bridge Between Science and Theology* (Downers Grove, IL: InterVarsity Press, 1999). J. A. Campbell and S. C. Meyer, *Darwinism, Design, and Public Education* (East Lansing: Michigan State University Press, 2003).

19. M. J. Behe, *The Edge of Evolution: The Search for the Limits of Darwinism* (New York: Free Press, 2007). W. A. Dembski and S. McDowell, *Understanding Intelligent Design* (Eugene, OR: Harvest House, 2008). S. C. Meyer, *Signature in the Cell: DNA and the Evidence for Intelligent Design* (New York: HarperOne, 2009). M. Behe, Self-organization and irreducibly complex systems: a reply to Shanks and Joplin. *Philosophy of Science 67*, no. 1 (2000): 155–62. W. A. Dembski, *No Free Lunch: Why Specified Complexity Cannot Be Purchased Without Intelligence* (Lanham, MD: Rowman & Littlefield, 2002).

20. M. Behe, Reply to my critics: a response to reviews of *Darwin's Black Box: The Biochemical Challenge to Evolution*. *Biology and Philosophy 16*, no. 5 (2001): 685–709. M. J. Behe, *The Edge of Evolution: The Search for the Limits of Darwinism* (New York: Free Press, 2007).

21. S. van der Linden, Why doctors should convey the medical consensus on vaccine safety. *Evidence-Based medicine 21*, no. 3 (2016): 119. J. S. Gerber and P. A. Offit, Vaccines and autism: a tale of shifting hypotheses. *Clinical Infectious Diseases 48*, no. 4 (2009): 456–61. G. A. Poland and R. M. Jacobson, The age-old struggle against the antivaccinationists. *New England Journal of Medicine 364*, no. 2 (2011): 97–99. P. A. Offit, *Deadly Choices: How the Anti-Vaccine Movement Threatens Us All* (New York: Basic Books, 2011). G. A. Poland, MMR vaccine and autism: Vaccine nihilism and postmodern science. *Mayo Clinic Proceedings 86*, no. 9 (2011): 869–71.

22. P. Hobson-West, "Trusting blindly can be the biggest risk of all": organised resistance to childhood vaccination in the UK. *Sociology of Health and Illness 29*, no. 2 (2007): 198–215. M. Holland et al., Unanswered questions from the Vaccine Injury Compensation Program: a review of compensated cases of vaccine-induced brain injury. *Pace Environmental Law Review 28*, no. 2 (2011): 480–543. M. Navin, *Values and Vaccine Refusal: Hard Questions in Ethics, Epistemology, and Health Care* (New York: Routledge, 2016).

23. M. J. Goldenberg, Public misunderstanding of science? Reframing the problem of vaccine hesitancy. *Perspectives on Science 24*, no. 5 (2016): 552–81.

24. Longino, *Science as Social Knowledge*. Longino, *The Fate of Knowledge*. Solomon, *Social Empiricism*.

25. Longino, *Science as Social Knowledge*. Longino, *The Fate of Knowledge*. Solomon, *Social Empiricism*. Kuhn, *The Essential Tension*.

26. Kuhn, *The Essential Tension*. L. Laudan, The epistemic, the cognitive, and the social, in P. K. Machamer and G. Wolters, eds., *Science, Values, and Objectivity* (Pittsburgh, PA: University of Pittsburgh Press, 2004), 14–23. H. Douglas, *Science, Policy, and the Value-Free Ideal* (Pittsburgh, PA: University of Pittsburgh Press, 2009). H. Douglas, The value of cognitive values, *Philosophy of Science 80*, no. 5 (2013): 796–806.

27. Longino, *Science as Social Knowledge*. J. A. Kourany, *Philosophy of Science after Feminism* (Oxford and New York: Oxford University Press, 2010).

28. N. Cartwright, Well-ordered science: evidence for use. *Philosophy of Science* 73, no. 5 (2006): 981–90. Douglas, *Science, Policy, and the Value-Free Ideal*.

29. Longino, *Science as Social Knowledge*. Borgerson, Amending and defending.

30. Longino, *The Fate of Knowledge*. Solomon, *Norms of Dissent*. Kitcher, *Living with Darwin*.

31. Borgerson, Amending and defending. Solomon 2001, Solomon, *Norms of Dissent*.

32. A. Wylie, A plurality of pluralisms: collaborative practice in archaeology, in F. Padovani, A. Richardson, and J. Y. Tsou, eds., *Objectivity in Science* (Dordrecht, NL: Springer, 2015), 189–210.

33. S. D. Mitchell, *Biological Complexity and Integrative Pluralism* (Cambridge, UK, and New York: Cambridge University Press, 2003).

34. Mitchell, *Biological Complexity and Integrative Pluralism*. J. Dupré, *The Disorder of Things: Metaphysical Foundations of the Disunity of Science* (Cambridge, MA: Harvard University Press, 1993). H. Longino, Gender, politics, and the theoretical virtues, *Synthese* 104, no. 3 (1995): 383–97.

35. A. Wylie, Community-based collaborative archaeology, in E. N. Cartwright, ed., *Philosophy of Social Science: A New Introduction* (New York: Oxford University Press, 2014), 68–82. Longino, *The Fate of Knowledge*.

36. Longino, Gender, politics, and the theoretical virtues. Kuhn, *The Essential Tension*. G. Doppelt, Kuhn's epistemological relativism: Interpretation and defense. *Inquiry–An Interdisciplinary Journal of Philosophy* 21, no. 1 (1978): 33–86. G. Doppelt, The value-ladenness of scientific knowledge, in J. Dupre, H. Kincaid, and A. Wylie, eds., *Value-Free Science? Ideals and Illusions* (Oxford: Oxford University Press, 2007), 188–217.

37. A. Wylie, Why standpoint matters, in R. Figueroa and S. Harding, eds., *Science and Other Cultures: Issues in Philosophies of Science and Technology* (New York: Routledge, 2003), 26–48.

38. E. Sober, *Reconstructing the Past: Parsimony, Evolution, and Inference* (Cambridge, MA: MIT Press, 1988). A. Plutynski, Parsimony and the Fisher-Wright debate, *Biology & Philosophy* 20, no. 4 (2005): 697–713. D. Steel, Testability and Ockham's razor: how formal and statistical learning theory converge in the new riddle of induction, *Journal of Philosophical Logic* 38, no. 5 (2009): 471–89. A. Zellner, H. A. Keuzenkamp, and M. McAleer, *Simplicity, Inference and Modeling: Keeping It Sophisticatedly Simple* (Cambridge, UK, and New York: Cambridge University Press, 2001).

39. Behe, *Darwin's Black Box*.

40. M. Ruse, *Darwinism Defended: A Guide to the Evolution Controversies* (Reading, MA: Addison-Wesley, 1982).

41. Longino, *Science as Social Knowledge*. Longino, *The Fate of Knowledge*. Borgerson, Amending and defending. B. Miller, When is consensus knowledge-based? Distinguishing shared knowledge from mere agreement, *Synthese* 190, no. 7 (2013): 1293–316.

42. S. Fall et al. Analysis of the impacts of station exposure on the US Historical Climatology Network temperatures and temperature trends, *Journal of Geophysical Research-Atmospheres* 116 (2011): D14120. B. Lomborg, *Cool It: The Skeptical Environmentalist's Guide To Global Warming* (New York: Vintage Books, 2008).

43. E. R. Wahl and C. M. Ammann, Robustness of the Mann, Bradley, Hughes reconstruction of Northern Hemisphere surface temperatures: examination of criticisms based on the nature and processing of proxy climate evidence, *Climatic Change* 85, no. 1–2 (2007): 33–69. Biddle and Leuschner, Climate skepticism and the manufacture of doubt. R. E. Benestad et al., Learning from mistakes in climate research, *Theoretical and Applied Climatology* 126, no. 3-4 (2015): 699–703.

44. Kitcher, *Science in a Democratic Society*.

45. L. DeFrancesco, How safe does transgenic food need to be? *Nature Biotechnology* 31, no. 9 (2013): 794–802. H. Lacey, *Values and Objectivity in Science* (Lanham, MD: Rowman & Littlefield, 2005).

46. Longino, *Science as Social Knowledge*.

47. S. L. Washburn and C. S. Lancaster, The evolution of hunting, in R. B. Lee and I. De Vore, eds., *Man the Hunter* (Chicago: Aldine, 1968), 293–303. W. S. Laughlin, Hunting: an integrating biobehavior system and its evolutionary importance, in R. B. Lee and I. DeVore, eds., *Man the Hunter* (Chicago: Aldine, 1968), 304–20. E. O. Wilson, *Sociobiology: The New Synthesis* (Cambridge, MA: Belknap Press of Harvard University Press, 1975).

48. A. Wylie and L. H. Nelson, Coming to terms with the values of science: insights from feminist science studies scholarship, in H. Kincaid et al., eds., *Value-Free Science? Ideals and Illusions* (Oxford: Oxford University Press, 2007), 58–86. F. Dahlberg, *Woman the Gatherer* (New Haven, CT: Yale University Press, 1981). J. Gero, Genderlithics: women's roles in stone tool production, in J. Gero and M. W. Conkey, eds., *Engendering Archaeology* (Oxford: Blackwell, 1991), 163–92. A. Wylie, Doing social science as a feminist: the engendering of archaeology, in A. N. H. Creager, E. Lunbeck, and L. Schiebinger, eds., *Feminism in Twentieth Century Science, Technology, and Medicine* (Chicago: University of Chicago Press, 2001), 23–45.

49. To be clear, our intention here is not to claim that it is not possible to recognize cases of flawed methodologies, problematic reasoning, or bias in scientific research. Appealing to shared epistemic norms is quite appropriate in determining whether we should or should not accept particular theories. Our concern here is with whether the notion of shared standards can be used as a criterion to *reliably* identify cases of NID. As we have said before, such reliability is important if an aim of identifying NID is to implement strategies that could in one way or another contribute to the silencing of epistemically valuable dissent.

50. Harker, *Creating Scientific Controversies*. Cook et al., Consensus on consensus. Oreskes, The scientific consensus on climate change. N. Oreskes and E. Conway, Defeating the merchants of doubt, *Nature* 465, no. 7299 (2010): 686–87.

51. Harker, *Creating Scientific Controversies*, 153; emphasis added.

52. K. Knorr-Cetina, *Epistemic Cultures: How the Sciences Make Knowledge* (Cambridge, MA: Harvard University Press, 1999). A. Goldman, Experts: Which ones should you trust? *Philosophy and Phenomenological Research* 63, no. 1 (2001): 85–110. D. Kennedy, Challenging expert rule: the politics of global governance, *Sydney Journal of International Law* 27 (2005): 5–28. S. Jasanoff, *States of Knowledge: The Co-Production of Science and Social Order* (London and New York: Routledge, 2004). S. Jasanoff, *Designs on Nature: Science and Democracy in Europe and the United States* (Princeton, NJ: Princeton University Press, 2005). E. Selinger and R. P. Crease, *The Philosophy of Expertise* (New York: Columbia University Press, 2006). H. M. Collins, *Are We All Scientific Experts Now?* (Cambridge, UK: Polity Press, 2014). H. M. Collins and R. Evans, The third wave

of science studies: studies of expertise and experience. *Social Studies of Science* 32, no. 2 (2002): 235–96. H. M. Collins and R. Evans, *Rethinking Expertise* (Chicago: University of Chicago Press, 2007). A. Irwin and B. Wynne, *Misunderstanding Science? The Public Reconstruction of Science and Technology* (Cambridge, UK, and New York: Cambridge University Press, 1996).

53. J. Cook et al., Quantifying the consensus on anthropogenic global warming in the scientific literature, *Environmental Research Letters* 8, no. 2 (2013): 024024. Cook et al., Consensus on consensus. N. Oreskes, Beyond the ivory tower: the scientific consensus on climate change, *Science* 306, no. 5702 (2004): 1686. W. R. L. Anderegg et al., Expert credibility in climate change. *Proceedings of the National Academy of Sciences USA* 107, no. 27 (2010): 12107–109.

54. Cook et al., Consensus on consensus. B. Verheggen et al., Scientists' views about attribution of global warming, *Environmental Science and Technology* 48, no. 16 (2014): 8963–71.

55. Harker, *Creating Scientific Controversies*, 156.

56. Cook et al., Consensus on consensus. Cook et al., Quantifying the consensus. Oreskes and Conway, Defeating the merchants of doubt.

57. Oreskes and Conway, Defeating the merchants of doubt. Oreskes and Conway, *Merchants of Doubt*.

58. S. C. Meyer, The origin of biological information and the higher taxonomic categories, *Proceedings of the Biological Society of Washington* 117, no. 2 (2004): 213–39. M. J. Behe and D. W. Snoke, Simulating evolution by gene duplication of protein features that require multiple amino acid residues. *Protein Science* 13, no. 10 (2004): 2651–64. M. J. Behe, Experimental evolution, loss-of-function mutations, and "The First Rule of Adaptive Evolution," *Quarterly Review of Biology* 85, no. 4 (2010): 419–45.

59. Wylie, A plurality of pluralisms.

60. M. J. S. Rudwick, *The Great Devonian Controversy: The Shaping of Scientific Knowledge Among Gentlemanly Specialists* (Chicago: University of Chicago Press, 1985).

61. Kitcher, *Science in a Democratic Society*. A. Nicolia et al., An overview of the last 10 years of genetically engineered crop safety research, *Critical Reviews in Biotechnology* 34, no. 1 (2014): 77–88. B. M. Maghari and A. M. Ardekani, Genetically modified foods and social concerns, *Avicenna Journal of Medical Biotechnology* 3, no. 3 (2011): 109–17. J. L. Domingo and J. G. Bordonaba, A literature review on the safety assessment of genetically modified plants, *Environment International* 37, no. 4 (2011): 734–42. C. Snell et al., Assessment of the health impact of GM plant diets in long-term and multigenerational animal feeding trials: a literature review, *Food and Chemical Toxicology* 50, no. 3–4 (2012): 1134–48. A. Dona and I. S. Arvanitoyannis, Health risks of genetically modified foods, *Critical Reviews in Food Science and Nutrition* 49, no. 2 (2009): 164–75. Lacey, *Values and Objectivity in Science*. A. Hilbeck et al., No scientific consensus on GMO safety, *Environmental Sciences Europe* 27, no. 4 (2015): https://doi.org/10.1186/s12302-014-0034-1.

62. S. G. Harding, *Objectivity and Diversity: Another Logic of Scientific Research* (Chicago: University of Chicago Press, 2015). N. J. Reo and K. P. Whyte, Hunting and morality as elements of traditional ecological knowledge. *Human Ecology* 40, no. 1 (2012): 15–27. L. L. Schiebinger, *Plants and Empire: Colonial Bioprospecting in the Atlantic World* (Cambridge, MA: Harvard University Press, 2004). P. Nadasdy, *Hunters and Bureaucrats: Power, Knowledge, and Aboriginal-State Relations in the Southwest Yukon*

(Vancouver: University of British Columbia Press, 2003). S. B. Brush and D. Stabinsky, *Valuing Local Knowledge: Indigenous People and Intellectual Property Rights* (Washington, DC: Island Press, 1996). V. D. Nazarea, *Ethnoecology: Situated Knowledge/Located Lives* (Tucson: University of Arizona Press, 1999). T. A. de Sousa Araujo et al., A new approach to study medicinal plants with tannins and flavonoids contents from the local knowledge, *Journal of Ethnopharmacology* 120, no. 1 (2008): 72–80. K. P. Whyte, On the role of traditional ecological knowledge as a collaborative concept: a philosophical study, *Ecological Processes* 2, no. 1 (2013): 1–12. B. Wynne, Sheep farming after Chernobyl: a case study in communicating scientific information, *Environment* 31, no. 2 (1989): 10–39. S. Epstein, *Impure Science: AIDS, Activism, and the Politics of Knowledge* (Berkeley: University of California Press, 1996).

63. Wynne, Sheep farming. B. Wynne, Misunderstood misunderstanding: social identities and public uptake of science, *Public Understanding of Science* 1 (1992): 281–304.

64. We discuss the role of contextual values in science more fully in chapter 9.

65. I. de Melo-Martín and K. Intemann, Can ethical reasoning contribute to better epidemiology? A case study in research on racial health disparities, *European Journal of Epidemiology* 22, no. 4 (2007): 215–21. J. Callicott, L. Crowder, and K. Mumford, Current normative concepts in conservation, *Conservation Biology* 13, no. 1 (1999): 22–35. K. Elliott, The ethical significance of language in the environmental sciences: case studies from pollution research, *Ethics, Place, & Environment* 12, no. 2 (2009): 157–73. J. Dupré, Fact and value, in H. Kincaid et al., eds., *Value-Free Science? Ideals and Illusions* (Oxford: Oxford University Press, 2007), 27–41. E. Anderson, Uses of value judgments in science: a general argument, with lessons from a case study of feminist research on divorce, *Hypatia* 19, no. 1 (2004): 1–24.

66. T. B. Mersha and T. Abebe, Self-reported race/ethnicity in the age of genomic research: its potential impact on understanding health disparities, *Human Genomics* 9, no. 1 (2015). https://doi.org/10.1186/s40246-014-0023-x. T. Caulfield et al., Race and ancestry in biomedical research: exploring the challenges, *Genome Medicine* 1 (2009): 8. G. Guo et al., Genetic bio-ancestry and social construction of racial classification in social surveys in the contemporary United States, *Demography* 51, no. 1 (2014): 141–72. D. R. Williams and M. Sternthal, Understanding racial-ethnic disparities in health: sociological contributions, *Journal of Health and Social Behavior* 51 (2010): S15–27. L. Gannett, Biogeographical ancestry and race, *Studies in History and Philosophy of Science Part C: Studies in History and Philosophy of Biological and Biomedical Sciences* 47 (2014): 173–84.

67. de Melo-Martín and Intemann, Can ethical reasoning contribute to better epidemiology?

68. F. Wickson and B. Wynne, Ethics of science for policy in the environmental governance of biotechnology: MON819 maize in Europe, *Ethics, Policy and Environment* 15, no. 3 (2012): 321–40. K. Intemann and I. de Melo-Martín, Social values and scientific evidence: the case of the HPV vaccines. *Biology and Philosophy* 25, no. 2 (2010): 203–13.

69. Longino, Gender, politics, and the theoretical virtues.

70. H. Douglas, *Science, Policy, and the Value-Free Ideal* (Pittsburgh, PA: University of Pittsburgh Press, 2009). K. Shrader-Frechette, *Risk and Rationality: Philosophical Foundations for Populist Reforms* (Berkeley: University of California Press, 1991).

71. J. Biddle and E. Winsberg, Value judgements and the estimation of uncertainty in climate modeling, in P. D. Magnus and J. Busch, eds., *New Waves in Philosophy of*

Science (Basingstoke, NH: Palgrave MacMillan, 2010), 172–97. E. Winsberg, Values and uncertainties in the predictions of global climate models, *Kennedy Institute of Ethics Journal* 22, no. 2 (2012): 111–37. R. H. Moss and S. H. Schneider, Uncertainties in the IPCC TAR: recommendations to lead authors for more consistent assessment and reporting, in R. Pachauri, T. Taniguchi, and K. Tanaka, eds., *Guidance Papers on the Cross Cutting Issues of the Third Assessment Report of the IPCC* (Geneva: World Meteorological Organization, 2000), 33–51. J. P. van der Sluijs, Uncertainty and dissent in climate risk assessment: a postnormal perspective, *Nature + Culture* 7, no. 2 (2012): 174–95.

72. Intergovernmental Panel on Climate Change (IPCC), *Climate Change 2013: The Physical Science Basis.* Working Group I contribution to the fifth assessment report (Geneva: Intergovernmental Panel on Climate Change, 2013). L. O. Mearns, Quantification of uncertainties of future climate change: Challenges and applications. *Philosophy of Science* 77, no. 5 (2010): 998–1011.

73. IPCC, *Climate Change 2013,* 3–9.

74. Douglas, *Science, Policy, and the Value-Free Ideal.* Biddle and Winsberg, Value judgements and the estimation of uncertainty. Winsberg, Values and uncertainties.

75. Collins and Evans, The third wave of science studies. Collins and Evans, *Rethinking Expertise.* Selinger and Crease, *The Philosophy of Expertise.* Wynne, Sheep farming after Chernobyl. Wynne, Misunderstood misunderstanding. B. Wynne, Creating public alienation: expert cultures of risk and ethics on GMOs, *Scientific Culture* 10, no. 4 (2001): 445–81.

76. C. J. Kirchhoff, M. C. Lemos, and S. Dessai Actionable knowledge for environmental decision making: broadening the usability of climate science, *Annual Review of Environment and Resources* 38 (2013): 393–414. P. Kloprogge and J. P. van der Sluijs, The inclusion of stakeholder knowledge and perspectives in integrated assessment of climate change, *Climatic Change* 75, no. 3 (2006): 359–89. S. Tang and S. Dessai, Usable science? The U.K. Climate projections 2009 and decision support for adaptation planning, *Weather, Climate, and Society* 4 (2012): 300–13.

77. Oreskes and Conway, Defeating the merchants of doubt.

78. K. vanden Heuvel, Jenny McCarthy's fear-mongering and the cult of false equivalence, *The Nation,* July 22, 2013.

79. Longino, *Science as Social Knowledge.* Longino, *The Fate of Knowledge.*

Chapter 5

1. J. Tollefson, 2015 declared the hottest year on record, *Nature* 529, no. 7587 (2016): 450.

2. J. T. Abatzoglou and A. P. Williams, Impact of anthropogenic climate change on wildfire across western U.S. forests, *Proceedings of the National Academy of Sciences USA* 113, no. 42 (2016): 11770–75.

3. W. Maslowski et al., The future of Arctic sea ice, *Annual Review of Earth and Planetary Sciences* 40 (2012): 625–54. J. E. Overland and M. Wang, When will the summer Arctic be nearly sea ice free? *Geophysical Research Letters* 40, no. 10 (2013): 2097–101.

4. M. E. Hauer, J. M. Evans, and D. R. Mishra, Millions projected to be at risk from sea-level rise in the continental United States, *Nature Climate Change* 6, no. 7 (2016): 691–95.

5. J. Biddle and A. Leuschner, Climate skepticism and the manufacture of doubt: can dissent in science be epistemically detrimental? *European Journal for Philosophy of Science* 5, no. 3 (2015): 261–78.

6. Biddle and Leuschner also understand "dissent" as including both the act of objecting to a widely held position in the scientific community and the act of promoting the production and dissemination of such objections.

7. Some caveats are in order here. Biddle and Leuschner maintain that the aims of their discussion regarding epistemically detrimental dissent are different from our own (J. Biddle, I. Kidd, and A. Leuschner, Epistemic corruption and manufactured doubt: the case of climate science, *Public Affairs Quarterly* 31, no. 3 [2017]: 165–88, and personal communications). This seems correct. Their main concern is to argue that some dissent—dissent that meets the conditions of the IndRA account—is likely to be epistemically detrimental in certain social contexts. They thus focus on a particular consequence of what they call "bad" dissent (Biddle and Leuschner, Climate skepticism and the manufacture of doubt). We agree with the claim that some dissent can be epistemically detrimental. Indeed, the motivation for the present project is the recognition—shared by many—that some scientific dissent can have negative social and epistemic effects. Also, while Biddle and Leuschner offer their IndRA as a criterion to distinguish between good and bad dissent—or, in our terms, between normatively appropriate and inappropriate dissent—our goal is to evaluate whether the criteria offered are successful. Although our aims are different, we believe the arguments we offer are relevant to their proposal. We take them to be asking the question: "When is dissent likely to result in adverse epistemic effects?" (or in our terms, "When is dissent normatively inappropriate?"); the answer they give is, "When it meets the IndRA conditions" (and provided that certain social context obtains). Hence, their proposal falls squarely within our aims—that is, it constitutes a criterion for identifying NID.

8. Biddle and Leuschner, Climate skepticism and the manufacture of doubt.

9. J. Cook et al., Consensus on consensus: a synthesis of consensus estimates on human-caused global warming, *Environmental Research Letters* 11, no. 4 (2016): 048002.

10. R. Brulle, Institutionalizing delay: foundation funding and the creation of U.S. climate change counter-movement organizations, *Climatic Change* 122, no. 4 (2014): 681–94.

11. Intergovernmental Panel on Climate Change (IPCC), *Climate Change 2014: Impacts, Adaptation, and Vulnerability*, Working Group II contribution to the fifth assessment report of the Intergovernmental Panel on Climate Change. 2014.

12. IPCC, *Climate Change 2014*.

13. M. E. Mann, R. S. Bradley, and M. K. Hughes, Global-scale temperature patterns and climate forcing over the past six centuries, *Nature* 392, no. 6678 (1998): 779–87.

14. S. McIntyre and R. McKitrick, Corrections to Mann et al. (1998) proxy database and Northern Hemisphere average temperature series, *Energy and Environment* 14 (2003): 751–71.

15. McIntyre and McKitrick, Corrections to Mann et al. S. McIntyre and R. McKitrick, Hockey sticks, principle components, and spurious significance, *Geophysical Research Letters* 32 (2005): L03710.

16. E. R. Wahl and C. M. Ammann, Robustness of the Mann, Bradley, Hughes reconstruction of Northern Hemisphere surface temperatures: examination of criticisms based on the nature and processing of proxy climate evidence, *Climatic Change* 85, no. 1–2 (2007): 33–69. S. Rutherford et al., Proxy-based Northern Hemisphere surface temperature

reconstructions: sensitivity to method, predictor network, target season, and target domain, *Journal of Climate* 18, no. 13 (2005): 2308–29.

17. P. Huybers, Comment on "Hockey sticks, principal components, and spurious significance" by S. McIntyre and R. McKitrick, *Geophysical Research Letters* 32, L20705 (2005): doi:10.1029/2005GL023395.

18. Biddle and Leuschner, Climate skepticism and the manufacture of doubt.

19. Wahl and Ammann, Robustness of the Mann, Bradley, Hughes reconstruction. S. A. Marcott et al., A reconstruction of regional and global temperature for the past 11,300 years, *Science* 339, no. 6124 (2013): 1198–201. T. J. Osborn and K. R. Briffa, The spatial extent of 20th-century warmth in the context of the past 1200 years, *Science* 311, no. 5762 (2006): 841–844. Rutherford et al., Proxy-based Northern Hemisphere surface temperature reconstructions.

20. Although we believe that taking into account the social and economic conditions under which modern science is produced today is important when investigating scientific dissent—or consensus, for that matter—the IndRA might suffer from excessively limiting contextual considerations that seem to apply predominantly in the United States. This creates two problems for the account. One is a concern about the generalizability of the IndRA. The contextual factors they contend make some dissent, such as climate change, likely to be epistemically detrimental are arguably factors that obtain in the United States but that might be less prevalent, or not at all, in other countries. If so, it is not clear that the IndRA could identify NID in those other countries. Second, but related, given that much of science today, and particularly climate change science, is produced globally, it is unclear how the IndRA can justify the authors' claim that climate change dissent results in adverse epistemic effects. After all, if the conditions under which such dissent is likely to be detrimental are present in the United States but, say, not in Europe or Asia, presumably scientists in Europe and Asia would be unaffected by the context of intimidation that allegedly hinders scientific progress. Indeed, scientific, social, and policy discussion regarding climate change in Europe and the United States are different, as are discussions related to, for instance, genetically modified products. We believe this presents a serious problem for Biddle and Leuschner's account of epistemically detrimental dissent. Nonetheless because our concern here is with the criteria to identify NID, we do not discuss such problems in detail. We assume here that one could provide broader contextual factors that might still allow us to use the IndRA to identify NID, even if in broadening such contextual factors fewer—or different—cases of dissent are so identified.

21. We discuss extensively the increasing commercialization of science in chapter 8.

22. Battelle Memorial Institute, *2014 Global R&D Funding Forecast* (Columbus, OH: Battelle, 2013).

23. Battelle Memorial Institute, *2014 Global R&D Funding Forecast*.

24. H. Moses et al., The anatomy of medical research U.S. and international comparisons, *Journal of the American Medical Association* 313, no. 2 (2015): 174–89.

25. S. Ehrhardt, L. J. Appel, and C. L. Meinert, Trends in National Institutes of Health funding for clinical trials registered in ClinicalTrials.gov, *Journal of the American Medical Association* 314, no. 23 (2015): 2566–67.

26. R. Proctor, *Golden Holocaust: Origins of the Cigarette Catastrophe and the Case for Abolition* (Berkeley: University of California Press, 2011). D. Michaels, *Doubt Is Their Product: How Industry's Assault on Science Threatens Your Health* (Oxford and

New York: Oxford University Press, 2008). N. Oreskes and E. M. Conway, *Merchants of Doubt: How a Handful of Scientists Obscured the Truth on Issues from Tobacco Smoke to Global Warming* (New York: Bloomsbury, 2010).

27. Coca-Cola's funding of health research and partnerships [editorial], *Lancet* 386, no. 10001 (2015): 1312.

28. A. O'Connor, Coke discloses millions in grants for health research and community programs, *New York Times*, September 22, 2015. https://well.blogs.nytimes.com/2015/09/22/coke-discloses-millions-in-grants-for-health-research-and-community-programs/.

29. Oreskes and Conway, *Merchants of Doubt*.

30. E.g., M. Carrier, Science in the grip of the economy, in M. Carrier et al., eds., *The Challenge of the Social and the Pressure of Practice: Science and Values Revisited* (Pittsburgh, PA: University of Pittsburgh Press, 2008), 217–34. S. Krimsky, *Science in the Private Interest: Has the Lure of Profits Corrupted Biomedical Research?* (Lanham, MD: Rowman & Littlefield, 2003). D. B. Resnik, *The Price of Truth: How Money Affects the Norms of Science* (Oxford and New York: Oxford University Press, 2007). J. R. Brown, Funding, objectivity and the socialization of medical research, *Science and Engineering Ethics* 8, no. 3 (2002): 295–308. K. Intemann and I. de Melo-Martín, Addressing problems in profit-driven research: How can feminist conceptions of objectivity help? *European Journal for Philosophy of Science* 4, no. 2 (2014): 135–51.

31. J. De Winter, How to make the research agenda in the health sciences less distorted, *Theoria* 27 (2012): 75–93. J. Reiss and P. Kitcher, Biomedical research, neglected diseases, and well-ordered science, *Theoria* 24 (2009): 263–82. M. Angell, *The Truth About the Drug Companies: How They Deceive Us and What to Do About It* (New York: Random House, 2004). Proctor, *Golden Holocaust*.

32. T. Wilholt, Bias and values in scientific research, *Studies in History and Philosophy of Science* 40, no. 1 (2009): 92–101. Krimsky, *Science in the Private Interest*. Resnik, *The Price of Truth*. K. Shrader-Frechette, Conceptual analysis and special-interest science: toxicology and the case of Edward Calabrese, *Synthese* 177, no. 3 (2010): 449–69.

33. L. Cosgrove et al., Under the influence: the interplay among industry, publishing, and drug regulation, *Accounts of Chemical Research* 23, no. 5 (2016): 257–79. S. Sismondo, Pharmaceutical company funding and its consequences: a qualitative systematic review, *Contemporary Clinical Trials* 29, no. 2 (2008): 109–13. D. E. Zinner et al., Participation of academic scientists in relationships with industry, *Health Affairs (Millwood)* 28, no. 6 (2009): 1814–25. R. Kneller et al., Industry-university collaborations in Canada, Japan, the UK and USA—with emphasis on publication freedom and managing the intellectual property lock-up problem, *PLOS One* 9, no. 3 (2014): e90302.

34. E. Negin, Internal documents show fossil fuel industry has been aware of climate change for decades, *Huffington Post*, July 8, 2015. www.huffingtonpost.com/elliott-negin/internal-documents-show-f_b_7749988.html.

35. Krimsky, *Science in the Private Interest*. Oreskes and Conway, *Merchants of Doubt*. Michaels, *Doubt Is Their Product*.

36. B. Martin, Suppression of dissent in science, in W. R. Freudenburg and T. I. K. Youn, eds., *Research in Social Problems and Public Policy* (Stamford, CT: JAI Press, 1999), 7:105–3510. E. Waltz, GM crops: battlefield, *Nature* 461, no. 7260 (2009): 27–32.

37. Martin, Suppression of dissent in science. Michaels, *Doubt Is Their Product*. Proctor, *Golden Holocaust*.

38. Biddle and Leuschner, Climate skepticism and the manufacture of doubt.

39. Biddle et al., Epistemic corruption and manufactured doubt.

40. P. Duesberg and D. Rasnick, The AIDS dilemma: drug diseases blamed on a passenger virus, *Genetica* 104, no. 2 (1998): 85–132.

41. A. J. Wakefield and S. M. Montgomery, Measles, mumps, rubella vaccine: through a glass, darkly, *Adverse Drug Reactions and Toxicological Reviews* 19, no. 4 (2000): 265–92.

42. P. Chigwedere and M. Essex, AIDS denialism and public health practice, *AIDS and Behavior* 14, no. 2 (2010): 237–47. M. McKee and P. Diethelm, How the growth of denialism undermines public health, *British Medical Journal* 341 (2010): c6950.

43. We discuss these concerns in more detail in chapter 8.

44. But see note 3, this chapter, for concerns about the contextual factors considerations.

45. Biddle and Leuschner, Climate skepticism and the manufacture of doubt, 274.

46. R. C. Jeffrey, Valuation and acceptance of scientific hypotheses, *Philosophy of Science* 23, no. 3 (1956): 237–46.

47. Biddle and Leuschner, Climate skepticism and the manufacture of doubt, 272.

48. Biddle and Leuschner, Climate skepticism and the manufacture of doubt, 272.

49. Biddle and Leuschner, Climate skepticism and the manufacture of doubt.

50. S. McIntyre and R. McKitrick, Reply to comment by Huybers on "Hockey sticks, principal components, and spurious significance," *Geophysical Research Letters* 32 (2005): 20.

51. M. W. Salzer et al., Changing climate response in near-treeline bristlecone pine with elevation and aspect, *Environmental Research Letters* 9, no. 11 (2014): 114007.

52. R. Pearce, Controversy behind climate science's "hockey stick" graph, *The Guardian,* February 2, 2010. https://www.theguardian.com/environment/2010/feb/02/hockey-stick-graph-climate-change.

53. L. Dawson et al., Considering usual medical care in clinical trial design, *PLOS Medicine* 6, no. 9 (2009): e1000111.

54. H. E. Longino, *The Fate of Knowledge* (Princeton, NJ: Princeton University Press, 2002). H. E. Longino, *Science as Social Knowledge: Values and Objectivity in Scientific Inquiry* (Princeton, NJ: Princeton University Press, 1990). T. S. Kuhn, *The Essential Tension: Selected Studies in Scientific Tradition and Change* (Chicago: University of Chicago Press, 1977). M. Solomon, *Social Empiricism* (Cambridge, MA: MIT Press, 2001).

55. Longino, *Science as Social Knowledge.* J. A. Kourany, *Philosophy of Science after Feminism* (Oxford and New York: Oxford University Press, 2010).

56. Indeed, Biddle and Leuschner explicitly recognize that the case of climate change is a complicated one and they acknowledge that determinations of who is affected by such risks are difficult. Yet, this constitutes a problem for their proposal to identify NID, at least if we wish such identification to be reliable.

Chapter 6

1. S. Lewandowsky, G. E. Gignac, and S. Vaughan, The pivotal role of perceived scientific consensus in acceptance of science, *Nature Climate Change* 3, no. 4 (2013): 399–404. M. Darby, UK public ignorant of climate science consensus—poll, *Climate Change News,* August 27, 2014. http://www.climatechangenews.com/2014/08/27/uk-public-ignorant-of-climate-science-consensus-poll/. Pew Research Center, Public and scientists' views on

science and society (2015). www.pewinternet.org/files/2015/01/PI_ScienceandSociety_ Report_012915.pdf.

2. Pew Research Center, Public and scientists' views.

3. J. Lewis and T. Speers, Misleading media reporting? The MMR story, *National Review of Immunology* 3, no. 11 (2003): 913–18.

4. See chapter 2 for discussion of these duties.

5. J. Biddle and A. Leuschner, Climate skepticism and the manufacture of doubt: can dissent in science be epistemically detrimental? *European Journal for Philosophy of Science* 5, no. 3 (2015): 261–78.

6. N. Oreskes and E. M. Conway, *Merchants of Doubt: How a Handful of Scientists Obscured the Truth on Issues from Tobacco Smoke to Global Warming* (New York: Bloomsbury, 2010). M. E. Mann, *The Hockey Stick and the Climate Wars: Dispatches from the Front Lines* (New York: Columbia University Press, 2012). Biddle and Leuschner, Climate skepticism.

7. A. Leuschner, Is it appropriate to "target" inappropriate dissent? On the normative consequences of climate skepticism, *Synthese* 195, no. 3 (2018):1255–1271. K. Brysse et al., Climate change prediction: erring on the side of least drama? *Global Environmental Change-Human and Policy Dimensions* 23, no. 1 (2013): 327–37. W. R. Freudenburg and V. Muselli, Global warming estimates, media expectations, and the asymmetry of scientific challenge, *Global Environmental Change-Human and Policy Dimensions* 20, no. 3 (2010): 483–91.

8. T. O. McGarity and W. Wagner, *Bending Science: How Special Interests Corrupt Public Health Research* (Cambridge, MA: Harvard University Pres, 2008). P. Diethelm and M. McKee, Denialism: what is it and how should scientists respond? *European Journal of Public Health* 19, no. 1 (2009): 2–4. M. McKee and P. Diethelm, How the growth of denialism undermines public health, *British Medical Journal* 341 (2010): c6950. Oreskes and Conway, *Merchants of Doubt*. M. Specter, The denialists: the dangerous attacks on the consensus about H.I.V. and AIDS, *New Yorker*, March 12, 2007. S. B. Omer et al., Vaccine refusal, mandatory immunization, and the risks of vaccine-preventable diseases, *New England Journal of Medicine* 360, no. 19 (2009): 1981–88.

9. P. Kitcher, *Science, Truth, and Democracy* (Oxford and New York: Oxford University Press, 2001). P. Kitcher, *The Advancement of Science: Science without Legend, Objectivity without Illusions* (New York: Oxford University Press, 1993). One might argue that this strategy could restrict the valuable roles that dissent can have, but recall that we are assuming here that criteria to reliably identify NID exist and thus, assuming that such criteria are used appropriately, there is no danger that epistemically valuable dissent could be limited.

10. R. Brulle, Institutionalizing delay: foundation funding and the creation of U.S. climate change counter-movement organizations, *Climatic Change* 122, no. 4 (2014): 681–94.

11. K. Stenius and T. F. Babor, The alcohol industry and public interest science, *Addiction* 105, no. 2 (2010): 191–98. E. Lipton, Food industry enlisted academics in G.M.O. lobbying war, emails show, *New York Times*, September 5, 2015. https://www.nytimes.com/2015/09/06/us/food-industry-enlisted-academics-in-gmo-lobbying-war-emails-show.html. A. O'Connor, Coke discloses millions in grants for health research and community programs, *New York Times,* September 22, 2015. https://well.blogs.nytimes.com/2015/09/22/coke-discloses-millions-in-grants-for-health-research-and-community-programs/.

12. Leuschner, Is it appropriate to target.

13. Biddle and Leuschner, Climate skepticism and the manufacture of doubt.

14. Oreskes and Conway, *Merchants of Doubt*. D. Michaels, *Doubt Is Their Product: How Industry's Assault on Science Threatens Your Health* (Oxford and New York: Oxford University Press, 2008). R. Proctor, *Golden Holocaust: Origins of the Cigarette Catastrophe and the Case for Abolition* (Berkeley: University of California Press, 2011). Mann, *The Hockey Stick*.

15. K. Intemann and I. de Melo-Martín, Are there limits to scientists' obligations to seek and engage dissenters? *Synthese* 191, no. 12 (2014): 2751–65.

16. P. Kitcher, *Science in a Democratic Society* (Amherst, NY: Prometheus, 2011).

17. P. Thacker, Skeptics get a journal, *Environmental Science and Technology Online*, August 31, 2005. http://www.realclimate.org/docs/thacker/skeptics.pdf.

18. Thacker, Skeptics get a journal.

19. J. M. Inhofe, *The Greatest Hoax: How the Global Warming Conspiracy Threatens Your Future* (Washington, DC: WND Books, 2012). S. McIntyre, *Climate Audit* (2016). https://climateaudit.org/. Friends of Science, President's message: FoS is dedicated to providing the public with insight into climate science. *FOS Membership Quarterly Newsletter* (2016). https://friendsofscience.org/assets/documents/2016_June_Newsletter.pdf.

20. National Vaccine Information Center (NVIC). (2016). www.nvic.org/. A. J. Wakefield, director, *Vaxxed: From Cover-Up to Catastrophe* (Produced by Del Bigtree, distributed by Cinema Libre Studio, 2016). L. K. Habakus and M. Holland, *Vaccine Epidemic: How Corporate Greed, Biased Science, and Coercive Government Threaten Our Human Rights, Our Health, and Our Children* (New York: Skyhorse, 2011).

21. We will discuss this case in more detail in chapter 9.

22. Q. Schiermeier, Storm clouds gather over leaked climate e-mails, *Nature* 462, no. 7272 (2009): 397.

23. National Oceanic and Atmospheric Administration (NOAA), Inspector General's review of stolen emails confirms no evidence of wrong-doing by NOAA climate scientists (February 25, 2011). www.noaanews.noaa.gov/stories2011/20110224_climate.html.

24. R. Grundmann, The legacy of climategate: revitalizing or undermining climate science and policy? *Wiley Interdisciplinary Reviews—Climate Change* 3, no. 3 (2012): 281–88. M. Ryghaug and T. M. Skjolsvold, The global warming of climate science: climategate and the construction of scientific facts, *International Studies in the Philosophy of Science* 24, no. 3 (2010): 287–307.

25. D. Harker, *Creating Scientific Controversies: Uncertainty and Bias in Science and Society* (Cambridge, UK: Cambridge University Press, 2015).

26. N. Oreskes and E. M. Conway, Defeating the merchants of doubt, *Nature* 465, no. 7299 (2010): 686–87. S. L. van der Linden et al., The scientific consensus on climate change as a gateway belief: experimental evidence, *PLOS One* 10, no. 2 (2015): e0118489. S. van der Linden, Why doctors should convey the medical consensus on vaccine safety, *Evidence-Based Medicine* 21, no. 3 (2016): 119.

27. J. Cook et al., Consensus on consensus: a synthesis of consensus estimates on human-caused global warming, *Environmental Research Letters* 11, no. 4 (2016): 048002.

28. N. Oreskes, Beyond the ivory tower: the scientific consensus on climate change, *Science* 306, no. 5702 (2004): 1686. W. R. L. Anderegg et al., Expert credibility in climate change, *Proceedings of the National Academy of Sciences USA* 107, no. 27 (2010): 12107–109. Cook et al., Consensus on consensus. J. Cook et al., Quantifying the consensus on anthropogenic global warming in the scientific literature, *Environmental Research Letters*

8, no. 2 (2013): 024024. B. Verheggen et al., Scientists' views about attribution of global warming, *Environmental Science and Technology* 48, no. 16 (2014): 8963–71. J. S. Carlton et al., The climate change consensus extends beyond climate scientists, *Environmental Research Letters* 10, no. 9 (2015): 094025.

29. Cook et al., Consensus on consensus.

30. D. Ding et al., Support for climate policy and societal action are linked to perceptions about scientific agreement, *Nature Climate Change* 1 (December 2011): 462–66. S. L. van der Linden et al., Highlighting consensus among medical scientists increases public support for vaccines: evidence from a randomized experiment, *BMC Public Health* 15 (2015): 1207. Lewandowsky et al., The role of conspiracist ideation. Lewandowsky et al., The pivotal role of perceived scientific consensus. A. M. McCright, R. E. Dunlap, and C. Xiao, Perceived scientific agreement and support for government action on climate change in the USA, *Climatic Change* 119, no. 2 (2013): 511–18.

31. McCright et al, Perceived scientific agreement. Ding et al., Support for climate policy.

32. Van der Linden et al., The scientific consensus on climate change. Lewandowsky et al., The role of conspiracist ideation. Lewandowsky et al., The pivotal role of perceived scientific consensus.

33. Van der Linden et al., The scientific consensus on climate change.

34. Van der Linden et al., Highlighting consensus among medical scientists.

35. Van der Linden et al., Highlighting consensus among medical scientists.

36. T. Deryugina and O. Shurchkov, The effect of information provision on public consensus about climate change, *PLOS One* 11, no. 4 (2016): e0151469.

37. D. M. Kahan, Climate-science communication and the measurement problem, *Political Psychology* 36 (2015): 1–43.

38. Kahan, Climate-science communication.

39. Kahan, Climate-science communication. D. M. Kahan et al., Cultural cognition of scientific consensus, *Journal of Risk Research* 14(2) (2011): 147–74.

40. Kahan et al., Cultural cognition of scientific consensus. A. Corner, L. Whitmarsh, and D. Xenias, Uncertainty, scepticism and attitudes towards climate change: biased assimilation and attitude polarization, *Climatic Change* 114, no. 3–4 (2012): 463–78.

41. J. Beatty, Masking disagreement among experts, *Episteme: A Journal of Social Epistemology* 3, no. 1 (2006): 52–67. M. Gilbert, Modelling collective belief, *Synthese* 73, no. 1 (1987): 185–204.

42. Beatty, Masking disagreement among experts. B. Miller, When is consensus knowledge-based? Distinguishing shared knowledge from mere agreement, *Synthese* 190, no. 7 (2013): 1293–316. M. Oppenheimer et al., Climate change: the limits of consensus, *Science* 317, no. 5844 (2007): 1505–506. J. P. van der Sluijs et al., Beyond consensus: reflections from a democratic perspective on the interaction between climate politics and science, *Current Opinion in Environmental Sustainability* 2, no. 5–6 (2010): 409–15.

43. Cook et al., Consensus on consensus. Cook et al., Quantifying the consensus on anthropogenic global warming.

44. Oppenheimer et al., Climate change: the limits of consensus. S. Lewandowsky et al., Seepage: climate change denial and its effect on the scientific community, *Global Environmental Change—Human and Policy Dimensions* 33 (2015): 1–13. R. S. Bradley, *Global Warming and Political Intimidation: How Politicians Cracked Down on Scientists*

as the Earth Heated Up (Amherst: University of Massachusetts Press, 2011). Brysse et al., Climate change prediction.

45. Van der Sluijs et al., Beyond consensus. J. van der Sluijs et al., Anchoring devices in science for policy: the case of consensus around climate sensitivity, *Social Studies of Science* 28, no. 2)(1998): 291–323.

46. E. Parasidis, Public health law and institutional vaccine skepticism, *Journal of Health and Political Policy Law* 41, no. 6 (2016): 1137–49.

47. I. de Melo-Martín and K. Intemann, Scientific dissent and public policy. Is targeting dissent a reasonable way to protect sound policy decisions? *EMBO Reports* 14, no. 3 (2013): 231–35. I. de Melo-Martín and K. Intemann, Who's afraid of dissent? Addressing concerns about undermining scientific consensus in public policy developments, *Perspective on Science* 22, no. 4 (2014): 593–615.

48. N. Pidgeon and B. Fischoff, The role of social and decisions sciences in communicating uncertain climate risks, *Nature Climate Change* 1 (2011): 35–41. Q. Schiermeier, The real holes in climate science, *Nature* 463, no. 7279 (2010): 284–87. D. J. Opel and E. K. Marcuse, The enigma of alternative childhood immunization schedules: what are the questions? *JAMA Pediatrics* 167, no. 3 (2013): 304–305.

49. Intergovernmental Panel on Climate Change (IPCC), *Climate Change 2007: Synthesis Report* (Geneva: Intergovernmental Panel on Climate Change, 2007). http://www.ipcc.ch/pdf/assessment-report/ar4/syr/ar4_syr.pdf. B. Weare, Insights into the importance of cloud vertical structure in climate, *Geophysical Research Letters* 27, no. 6 (2000): 907–10. B. Weare, Near-global observations of low clouds, *Journal of Climate* 13, no. 7 (2000): 1255–68. D. Rind et al., The way forward, in M. Chin, R. Kahn, and S. Schwartz, eds., *Atmospheric Aerosol Properties and Climate Impacts* (Washington, DC: National Aeronautics and Space Administration, 2009), 85–90.

50. Brysse et al., Climate change prediction. Oppenheimer et al., Climate change: the limits of consensus.

51. J. M. Glanz et al., A population-based cohort study of undervaccination in 8 managed care organizations across the United States, *JAMA Pediatrics* 167, no. 3 (2013): 274–81.

52. Opel and Marcuse, The enigma of alternative childhood immunization schedules.

53. Opel and Marcuse, The enigma of alternative childhood immunization schedules. Glanz et al., A population-based cohort study of undervaccination.

54. de Melo-Martín and Intemann, Scientific dissent and public policy.

55. Oreskes and Conway, *Merchants of Doubt*. Cook et al., Consensus on consensus.

56. Oreskes and Conway, *Merchants of Doubt*. K. Shrader-Frechette, Conceptual analysis and special-interest science: toxicology and the case of Edward Calabrese, *Synthese* 177, no. 3 (2010): 449–69. K. Elliott, *Is a Little Pollution Good for You? Incorporating Societal Values in Environmental Research* (New York: Oxford University Press, 2011).

57. E. Waltz, GM crops: battlefield, *Nature* 461, no. 7260 (2009): 27–32.

58. J. Schnittker and G. Karandinos, Methuselah's medicine: pharmaceutical innovation and mortality in the United States, 1960–2000, *Social Science & Medicine* 70, no. 7 (2010): 961–68. D. Smith and B. MacFadyen, Industry relationships between physicians and professional medical associations: corrupt or essential? *Surgical Endoscopy* 24, no. 2 (2010): 251–53. B. Zycher et al., Private sector contributions to pharmaceutical science: thirty-five summary case histories, *American Journal of Therapeutics* 17(1)

(2010): 101–20. L. M. Portilla and M. L. Rohrbaugh, Leveraging public private partnerships to innovate under challenging budget times, *Current Topics in Medicinal Chemistry* 14, no. 3 (2014): 326–29.

59. L. Frank, Scientific conduct. Charges don't stick to the Skeptical Environmentalist, *Science* 303, no. 5654 (2004): 28.

60. C. J. Lee et al., Bias in peer review, *Journal of the American Society for Information Science and Technology* 64, no. 1 (2013): 2–17.

Chapter 7

1. Pew Research Center, Public and scientists' views on science and society (2015). www.pewinternet.org/files/2015/01/PI_ScienceandSociety_Report_012915.pdf.

2. Pew, Public and scientists' views, 6.

3. Pew, Public and scientists' views, 43.

4. Pew, Public and scientists' views, 47.

5. Pew, Public and scientists' views, 44.

6. Pew, Public and scientists' views, 50.

7. Scientific American, In science we trust, *Scientific American* 303, no. 4 (2010): 56–59.

8. Ipsos Group, *Global Trends 2014*. http://www.ipsosglobaltrends.com/files/gts_2014_web.pdf.

9. Pew Research Center, Religion in Latin America: widespread change in a historically Catholic region (2014). www.pewforum.org/2014/11/13/religion-in-latin-america/.

10. J. Cook and P. Jacobs, Scientists are from Mars, laypeople are from Venus: an evidence-based rationale for communicating the consensus on climate, *Reports of the National Center for Science Education* 34, no. 6 (2014): 3.1–3.10.

11. S. Van der Linden, Why doctors should convey the medical consensus on vaccine safety, *Evidence-Based Medicine* 21, no. 3 (2016): 119. S. L. van der Linden et al., The scientific consensus on climate change as a gateway belief: experimental evidence, *PLOS One* 10, no. 2 (2015): e0118489. D. Ding et al., Support for climate policy and societal action are linked to perceptions about scientific agreement, *Nature Climate Change* 1 (December 2011): 462–66. M. Aklin and J. Urpelainen, Perceptions of scientific dissent undermine public support for environmental policy, *Environmental Science and Policy* 38 (2014): 173–77. A. M. McCright, R. E. Dunlap, and C. Xiao, Perceived scientific agreement and support for government action on climate change in the USA, *Climatic Change* 119, no. 2 (2013): 511–18.

12. B. R. McFadden, Examining the gap between science and public opinion about genetically modified food and global warming, *PLOS One* 11, no. 11 (2016): e0166140. M. A. Ranney and D. Clark, Climate change conceptual change: scientific information can transform attitudes, *Topics in Cognitive Science* 8(1) (2016): 49–75. R. E. Dunlap and A. M. McCright, A widening gap—Republican and Democratic views on climate change, *Environment* 50, no. 5 (2008): 26–35. K. S. Fielding and M. J. Hornsey, A social identity analysis of climate change and environmental attitudes and behaviors: insights and opportunities, *Frontiers in Psychology* 7, no. 121 (2016): doi.org/10.3389/fpsyg.2016.00121. A. Zia and A. M. Todd, Evaluating the effects of ideology on public understanding of climate change science: how to improve communication across ideological divides? *Public Understanding of Science* 19, no. 6 (2010): 743–61. D. M. Kahan, Ideology, motivated reasoning, and cognitive reflection, *Judgment and Decision Making* 8, no. 4 (2013): 407–24.

13. A. Corner et al., Uncertainty, scepticism and attitudes towards climate change: biased assimilation and attitude polarization, *Climatic Change* 114, no. 3–4 (2012): 463–78. B. R. McFadden and J. L. Lusk, Cognitive biases in the assimilation of scientific information on global warming and genetically modified food, *Food Policy* 54 (2015): 35–43. Kahan, Ideology.

14. D. Harker, *Creating Scientific Controversies: Uncertainty and Bias in Science and Society* (Cambridge, UK: Cambridge University Press, 2015). J. Biddle and A. Leuschner, Climate skepticism and the manufacture of doubt: can dissent in science be epistemically detrimental? *European Journal for Philosophy of Science* 5, no. 3 (2015): 261–78. N. Oreskes and E. M. Conway, Defeating the merchants of doubt, *Nature* 465, no. 7299 (2010): 686–87. T. O. McGarity and W. Wagner, *Bending Science: How Special Interests Corrupt Public Health Research* (Cambridge, MA: Harvard University Press, 2008).

15. S. Wagenknecht, Facing the incompleteness of epistemic trust: managing dependence in scientific practice, *Social Epistemology* 29, no. 2 (2015): 160–84.

16. J. Hardwig, Epistemic dependence, *Journal of Philosophy* 82, no. 7 (1985): 335–49. J. Hardwig, The role of trust in knowledge, *Journal of Philosophy* 88, no. 12 (1991): 693–708. T. Wilholt, Epistemic trust in science, *British Journal for the Philosophy of Science* 64, no. 2 (2013): 233–53. P. Kitcher, *The Advancement of Science: Science without Legend, Objectivity without Illusions* (New York: Oxford University Press, 1993). K. Frost-Arnold, Moral trust and scientific collaboration, *Studies in History and Philosophy of Science* 44, no. 3 (2013): 301–10.

17. H. Andersen, Collaboration, interdisciplinarity, and the epistemology of contemporary science, *Studies in History and Philosophy of Science* 56 (2016): 1–10.

18. N. Scheman, Epistemology resuscitated: objectivity and trustworthiness, in N. Tuana and S. Morgen, eds., *Engendering Rationalities* (Albany: State University of New York Press, 2001), 23–52. Wilholt, Epistemic trust in science. H. E. Grasswick, Scientific and lay communities: earning epistemic trust through knowledge sharing, *Synthese* 177, no. 3 (2010): 387–409. E. Anderson, Democracy, public policy, and lay assessments of scientific testimony, *Episteme: A Journal of Individual and Social Epistemology* 8, no. 2 (2011): 144–64.

19. A. Baier, Trust and antitrust, *Ethics* 96, no. 2 (1986): 231–60. D. Gambetta, *Trust: Making and Breaking Cooperative Relations* (New York: Blackwell, 1988). M. Hollis, *Trust within Reason* (Cambridge, UK, and New York: Cambridge University Press, 1998). K. Jones, Trust as an affective attitude, *Ethics* 107, no. 1 (1996): 4–25. R. Holton, Deciding to trust, coming to believe, *Australasian Journal of Philosophy* 72, no. 1 (1994): 63–76.

20. R. Hardin, *Trust and Trustworthiness* (New York: Russell Sage Foundation, 2002). N. N. Potter, *How Can I Be Trusted? A Virtue Theory of Trustworthiness* (Lanham, MD: Rowman & Littlefield, 2002). N. Daukas, Epistemic trust and social location, *Episteme: A Journal of Social Epistemology* 3, no. 1–2 (2006): 109–24.

21. Hardwig, The role of trust in knowledge. D. B. Resnik, Scientific research and the public trust, *Science and Engineering Ethics* 17, no. 3 (2011): 399–409.

22. M. Fricker, *Epistemic Injustice: Power and the Ethics of Knowing* (Oxford and New York: Oxford University Press, 2007).

23. Fricker, Epistemic Injustice. Daukas, Epistemic trust and social location. Scheman, Epistemology resuscitated. K. Dotson, A cautionary tale on limiting epistemic oppression, *Frontiers—A Journal of Women Studies* 33, no. 1 (2012): 24–47. G. Marsh, Trust, testimony,

and prejudice in the credibility economy, *Hypatia—A Journal of Feminist Philosophy* 26, no. 2 (2011): 280–93.

24. K. Rolin, Gender and trust in science, *Hypatia* 17, no. 4 (2002): 95–118.

25. Scheman, Epistemology resuscitated. Grasswick, Scientific and lay communities. Rolin, Gender and trust.

26. Harker, *Creating Scientific Controversies*. Biddle and Leuschner, Climate skepticism and the manufacture of doubt. Oreskes and Conway, Defeating the merchants of doubt.

27. M. Specter, The denialists: the dangerous attacks on the consensus about H.I.V. and AIDS, *New Yorker*, March 12, 2007. M. Weinel, Primary source knowledge and technical decision-making: Mbeki and the AZT debate, *Studies in History and Philosophy of Science* 38, no. 4 (2007): 748–60. N. Nattrass, AIDS and the scientific governance of medicine in post-apartheid South Africa, *African Affairs* 107, no. 427 (2008): 157–76. N. Simelela et al., A political and social history of HIV in South Africa, *Current HIV/AIDS Reports* 12, no. 2 (2015): 256–61.

28. P. Chigwedere and M. Essex, AIDS denialism and public health practice, *AIDS and Behavior* 14, no. 2 (2010): 237–47.

29. P. Chigwedere et al., Estimating the lost benefits of antiretroviral drug use in South Africa, *Journal of Acquired Immune Deficiency Syndrome* 49, no. 4 (2008): 410–15. Nattrass, AIDS and the scientific governance of medicine.

30. Chigwedere et al., Estimating the lost benefits. Nattrass, AIDS and the scientific governance of medicine.

31. H. E. Purkitt and S. F. Burgess, *South Africa's Weapons of Mass Destruction* (Bloomington: Indiana University Press, 2005). C. Kenyon, Cognitive dissonance as an explanation of the genesis, evolution and persistence of Thabo Mbeki's HIV denialism, *African Journal of Aids Research* 7, no. 1 (2008): 29–35.

32. M. Gevisser, *Thabo Mbeki: The Dream Deferred* (Johannesburg: Jonathan Ball, 2007).

33. T. Mbeki, Address by President Thabo Mbeki, at the inaugural ZK Matthews Memorial Lecture, University of Fort Hare (2001). https://www.nelsonmandela.org/omalley/index.php/site/q/03lv03445/04lv04206/05lv04302/06lv04303/07lv04304.htm.

34. M. Mbali, AIDS discourses and the South African state: Government denialism and post-apartheid AIDS policy-making, *Transformation: Critical Perspectives on Southern Africa* 54, no. 1 (2004): 104–22. Gevisser, *Thabo Mbeki*. Kenyon, Cognitive dissonance.

35. N. A. Christakis, The ethical design of an AIDS vaccine trial in Africa, *Hastings Center Report* 18, no. 3 (1988): 31–37.

36. Kenyon, Cognitive dissonance.

37. Gevisser, *Thabo Mbeki*.

38. Rasmussen Reports, 69% say it's likely that scientists have falsified global warming research (February 1, 2011). www.rasmussenreports.com/public_content/politics/current_events/environment_energy/69_say_it_s_likely_scientists_have_falsified_global_warming_research.

39. Public Policy Polling, Conspiracy theory poll results (February 1, 2013). www.publicpolicypolling.com/main/2013/04/conspiracy-theory-poll-results-.html.

40. Pew Research Center, The politics of climate (2016). www.pewinternet.org/2016/10/04/the-politics-of-climate/.

41. S. McIntyre and R. McKitrick, Corrections to Mann et al. (1998) proxy database and Northern Hemisphere average temperature series, *Energy and Environment* 14 (2003): 751–71. S. McIntyre and R. McKitrick, Hockey sticks, principal components, and spurious significance, *Geophysical Research Letters* 32 (2005): L03710.

42. P. Huybers, Comment on "Hockey sticks, principal components, and spurious significance" by S. McIntyre and R. McKitrick, *Geophysical Research Letters* 32, no. 20 L20705 (2005): doi:10.1029/2005GL023395. E. R. Wahl and C. M. Ammann, Robustness of the Mann, Bradley, Hughes reconstruction of Northern Hemisphere surface temperatures: examination of criticisms based on the nature and processing of proxy climate evidence, *Climatic Change* 85, no. 1–2 (2007): 33–69.

43. S. Krimsky. An Illusory Consensus behind GMO Health Assessment. *Science Technology & Human Values, 40*, no. 6 (2015): 883–914.

44. M. J. Goldenberg, Public misunderstanding of science? Reframing the problem of vaccine hesitancy, *Perspectives on Science* 24, no. 5 (2016): 552–81.

Chapter 8

1. H. Ibsen, *An Enemy of the People* (Mineola, NY: Dover, [1882] 1999).

2. N. Scheman, Epistemology resuscitated: objectivity and trustworthiness, in N. Tuana and S. Morgen, eds., *Engendering Rationalities* (Albany: State University of New York Press, 2001), 23–52.

3. Scheman, Epistemology resuscitated. H. E. Grasswick, Scientific and lay communities: earning epistemic trust through knowledge sharing, *Synthese* 177, no. 3 (2010): 387–409. K. Rolin, Gender and trust in science, *Hypatia* 17, no. 4 (2002): 95–118. M. Fricker, *Epistemic Injustice: Power and the Ethics of Knowing* (Oxford and New York: Oxford University Press, 2007).

4. J. H. Jones, *Bad Blood: The Tuskegee Syphilis Experiment* (New York: Free Press, 1981). H. A. Washington, *Medical Apartheid: The Dark History of Medical Experimentation on Black Americans from Colonial Times to the Present* (New York: Harlem Moon, 2006). F. S. Hodge, No meaningful apology for American Indian unethical research abuses, *Ethics and Behavior* 22, no. 6 (2012): 431–44.

5. R. Dresser, Wanted: single, white male for medical research, *The Hastings Center Report* 22, no. 1 (1992): 24–29. V. Simon, Wanted: women in clinical trials, *Science* 308, no. 5728 (2005): 1517. A. C. Mastroianni et al., *Women and Health Research: Ethical and Legal Issues of Including Women in Clinical Studies* (Washington, DC: National Academies Press, 1994).

6. R. Jordan-Young and R. I. Rumiati, Hardwired for sexism? Approaches to sex/gender in neuroscience, *Neuroethics* 5, no. 3 (2012): 305–15. A. Fausto-Sterling, *Myths of Gender: Biological Theories about Women and Men* (New York: Basic Books, 1992). H. Rose and S. P. R. Rose, *Alas, Poor Darwin: Arguments against Evolutionary Psychology* (New York: Harmony Books, 2000).

7. K. S. Shrader-Frechette, *Burying Uncertainty: Risk and the Case against Geological Disposal of Nuclear Waste* (Berkeley: University of California Press, 1993). C. A. Zimring, *Clean and White: A History of Environmental Racism in the United States* (New York: New York University Press, 2015).

8. R. W. Durant et al., Different types of distrust in clinical research among whites and African Americans, *Journal of the National Medical Association* 103, no. 2 (2011): 123–30.

D. P. Scharff et al., More than Tuskegee: understanding mistrust about research participation, *Journal of Health Care for the Poor and Underserved* 21, no. 3 (2010): 879–97. R. M. Jordan-Young, *Brain Storm: The Flaws in the Science of Sex Differences* (Cambridge, MA: Harvard University Press, 2010). As discussed in chapter 7, trust is a complex, multifactorial phenomenon. Hence, to say that certain communities tend to be suspicious of scientists and their claims does not mean that such distrust or lack of trust extends to all claims or to all scientific disciplines equally. Thus, African Americans might be particularly suspicious of clinical research but not necessarily of research in the physical sciences. Similarly, women might be suspicious of claims from the biological sciences but perhaps not so of those from the environmental ones.

9. This does not mean that everyone's trust will be affected to the same degree.

10. B. A. Nosek and T. M. Errington, Making sense of replications, *Elife* 6 (2017): e23383. C. G. Begley and J. P. Ioannidis, Reproducibility in science: improving the standard for basic and preclinical research, *Circulation Research* 116, no. 1 (2015): 116–26. F. S. Collins and L. A. Tabak, NIH plans to enhance reproducibility, *Nature* 505, no. 7485 (2014): 612–13. E. Dolgin, Drug discoverers chart path to tackling data irreproducibility, *National Reviews Drug Discovery* 13 (2014): 875–76. S. Perrin, Preclinical research: make mouse studies work, *Nature* 507, no. 7493 (2014): 423–25. F. Prinz , T. Schlange, and K. Asadullah, Believe it or not: how much can we rely on published data on potential drug targets? *Nature Reviews Drug Discovery* 10, no. 9 (2011): 712.

11. T. Caulfield and U. Ogbogu, The commercialization of university-based research: balancing risks and benefits, *BMC Medical Ethics* 16 (2015): 70.

12. National Science Board, *Science and Engineering Indicators 2016* (Arlington, VA: National Sciences Foundation, 2016).

13. R. Wuestenhagen and E. Menichetti, Strategic choices for renewable energy investment: conceptual framework and opportunities for further research, *Energy Policy* 40 (2012): 1–10.

14. H. Moses et al., The anatomy of medical research U.S. and international comparisons, *Journal of the American Medical Association* 313, no. 2 (2015): 174–89.

15. Moses et al., The anatomy of medical research.

16. K. O. Fuglie and A. A. Toole, The evolving institutional structure of public and private agricultural research, *American Journal of Agricultural Economics* 96, no. 3 (2014): 862–83.

17. National Science Board, *Science and Engineering Indicators 2016*.

18. S. Ehrhardt et al., Trends in National Institutes of Health funding for clinical trials registered in ClinicalTrials.gov, *Journal of the American Medical Association* 314, no. 23 (2015): 2566–67.

19. National Science Board, *Science and Engineering Indicators 2016*.

20. National Science Board, *Science and Engineering Indicators 2016*.

21. V. Loise and A. J. Stevens, The Bayh-Dole Act turns 30, *Science Translational Medicine* 2, no. 52 (2010): 52cm27.

22. F. Lissoni, Academic patenting in Europe: a reassessment of evidence and research practices, *Industry and Innovation* 20, no. 5 (2013): 379–84. G. B. Wang and J. C. Guan, The role of patenting activity for scientific research: a study of academic inventors from China's nanotechnology, *Journal of Informetrics* 4, no. 3 (2010): 338–50. M. Takahashi and R. Carraz, Academic patenting in Japan: illustration from a leading Japanese university, in

Poh Kam Wong, ed., *Academic Entrepreneurship in Asia: The Role and Impact of Universities in National Innovation Systems* (Cheltenham: Edward Elgar, 2011), 86–107. R. Kneller et al., Industry–university collaborations in Canada, Japan, the UK and USA—with emphasis on publication freedom and managing the intellectual property lock-up problem, *PLOS One* 9, no. 3 (2014): e90302.

23. P. Kitcher, *Science, Truth, and Democracy* (Oxford and New York: Oxford University Press, 2001). J. Dupré, Fact and value, in H. Kincaid et al., eds., *Value-Free Science? Ideals and Illusions* (Oxford: Oxford University Press, 2007), 27–41. E. Anderson, Knowledge, human interests, and objectivity in feminist epistemology, *Philosophical Topics* 23, no. 2 (1995): 27–58. Scheman, Epistemology resuscitated.

24. Scheman, Epistemology resuscitated. Anderson, Knowledge, human interests. Grasswick, Scientific and lay communities.

25. F. Hendriks, D. Kienhues, and R. Bromme, Trust in science and the science of trust, in B. Blöbaum, ed., *Trust and Communication in a Digitized World* (Dordercht, NL: Springer, 2016), 143–59.

26. J. Hardwig, The role of trust in knowledge, *Journal of Philosophy* 88, no. 12 (1991): 693–708. D. B. Resnik, Scientific research and the public trust, *Science and Engineering Ethics* 17, no. 3 (2011): 399–409.

27. T. Pogge, The Health Impact Fund and its justification by appeal to human rights, *Journal of Social Philosophy* 40, no. 4 (2009): 542–69. J. Reiss and P. Kitcher, Biomedical research, neglected diseases, and well-ordered science, *Theoria* 24 (2009): 263–82. Caulfield and Ogbogu, The commercialization of university-based research. J. De Winter, How to make the research agenda in the health sciences less distorted, *Theoria* 27 (2012): 75–93.

28. Moses et al., The anatomy of medical research.

29. Moses et al., The anatomy of medical research.

30. J. L. Dieleman et al., Global health development assistance remained steady in 2013 but did not align with recipients' disease burden, *Health Affairs* 33, no. 5 (2014): 878–86.

31. Moses et al., The anatomy of medical research.

32. M. Angell, *The Truth about the Drug Companies: How They Deceive Us and What to Do about It* (New York: Random House, 2004). S. Krimsky, *Science in the Private Interest: Has the Lure of Profits Corrupted Biomedical Research?* (Lanham, MD: Rowman & Littlefield, 2003). T. O. McGarity and W. Wagner, *Bending Science: How Special Interests Corrupt Public Health Research* (Cambridge, MA: Harvard University Press, 2008).

33. K. Intemann and I. de Melo-Martín, Social values and scientific evidence: the case of the HPV vaccines, *Biology and Philosophy* 25, no. 2 (2010): 203–13.

34. Angell, *The Truth about the Drug Companies*. Krimsky, *Science in the Private Interest*. H. Brody, *Hooked: Ethics, the Medical Profession, and the Pharmaceutical Industry* (Lanham, MD: Rowman & Littlefield, 2007).

35. R. Moodie et al., Profits and pandemics: prevention of harmful effects of tobacco, alcohol, and ultra-processed food and drink industries, *Lancet* 381, no. 9867 (2013): 670–79.

36. Moses et al., The anatomy of medical research. R. S. Magnusson, Rethinking global health challenges: towards a "global compact" for reducing the burden of chronic disease, *Public Health* 123, no. 3 (2009): 265–74. J. E. James, Personalised medicine, disease prevention, and the inverse care law: more harm than benefit? *European Journal of Epidemiology* 29, no. 6 (2014): 383–90. T. Caulfield, Obesity genes, personalized medicine, and public health policy, *Current Obesity Reports* 4, no. 3 (2015): 319–23. E. T. Juengst

et al., Serving epigenetics before its time, *Trends in Genetics* 30, no. 10 (2014): 427–29. S. H. Woolf et al., Giving everyone the health of the educated: an examination of whether social change would save more lives than medical advances, *American Journal of Public Health* 97, no. 4 (2007): 679–83. G. P. Mays and S. A. Smith, Evidence links increases in public health spending to declines in preventable deaths, *Health Affairs* 30, no. 8 (2011): 1585–93. S. Chapman and R. MacKenzie, The global research neglect of unassisted smoking cessation: causes and consequences, *PLOS Medicine* 7, no. 2 (2010): e1000216.

37. Fuglie and Toole, The evolving institutional structure.

38. Wuestenhagen and Menichetti, Strategic choices.

39. W. L. Kilama, The 10/90 gap in sub-Saharan Africa: resolving inequities in health research, *Acta Tropica* 112, Suppl. 1 (2009): S8–15. P. Stevens, ed., *Fighting the Diseases of Poverty* (New Brunswick, NJ: Transaction, 2008). Pogge, The Health Impact Fund. Reiss and Kitcher, Biomedical research. S. H. E. Harmon, In search of global health justice: a need to reinvigorate institutions and make international law, *Health Care Analysis* 23, no. 4 (2015): 352–75. De Winter, How to make the research agenda.

40. We recognize that polls measuring public trust in science and other social institutions may be problematic. Insofar as trust is a complex phenomenon, what exactly polls measure when they inquire about trust is not always clear. In general, polls do not distinguish between trustworthiness and credibility, and as we discussed in chapter 7, such is a distinction is important to determine whether people's trust or distrust is warranted or not. For an author skeptic about polls measuring public trust, see O. O'Neill, *A Question of Trust* (Cambridge, UK: Cambridge University Press, 2002). Nonetheless, they provide some evidence that commercialization is impacting people's attitudes about science in negative ways.

41. Pew Research Center, Public and scientists' views on science and society (2015), 13. www.pewinternet.org/files/2015/01/PI_ScienceandSociety_Report_012915.pdf.

42. J. C. Besley et al., Perceived conflict of interest in health science partnerships, *PLOS One* 12, no. 4 (2017): e0175643. C. R. Critchley, Public opinion and trust in scientists: the role of the research context, and the perceived motivation of stem cell researchers, *Public Understanding of Science* 17, no. 3 (2008): 309–27. Hendriks et al., Trust in science. C. R. Critchley and D. Nicol, Understanding the impact of commercialization on public support for scientific research: is it about the funding source or the organization conducting the research? *Public Understanding of Science* 20, no. 3 (2011): 347–66.

43. M. Kessel, Restoring the pharmaceutical industry's reputation [commentary], *Nature Biotechnology* 32, no. 10 (2014): 983–90.

44. A. Banerjee, T. Pogge, and A. Hollis, The Health Impact Fund, *Lancet* 375, no. 9727 (2010): 166–9. T. Pogge. The Health Impact Fund: Enhancing Justice and Efficiency in Global Health. *Journal of Human Development and Capabilities*, 13, no. 4 (2012): 537–559.

45. J. R. Brown, The community of science, in M. Carrier et al., eds., *The Challenge of the Social and the Pressure of Practice: Science and Values Revisited* (Pittsburgh, PA: University of Pittsburgh Press, 2008), 189–216.

46. Reiss and Kitcher, Biomedical research.

47. De Winter, How to make the research agenda.

48. W. D. Valdivia, The stakes in Bayh-Dole: public values beyond the pace of innovation, *Minerva* 49, no. 1 (2011): 25–46. B. Collinsworth and S. E. Crager, Should academic therapeutic patents go to the highest bidder? *Expert Opinion on Therapeutic Patents*

24, no. 5 (2014): 481–84. S. Loewenberg, The Bayh-Dole Act: a model for promoting research translation? *Molecular Oncology* 3, no. 2 (2009): 91–93. R. L. Geiger and C. M. Sá, *Tapping the Riches of Science: Universities and the Promise of Economic Growth* (Cambridge, MA: Harvard University Press, 2008).

49. World Association of Medical Editors, Conflict of interest in peer-reviewed medical journals: a policy statement of the World Association of Medical Editors (WAME), *Journal of Child Neurology* 24, no. 10 (2009): 1321–23. Institute of Medicine, *Conflict of Interest in Medical Research, Education, and Practice* (Washington, DC: National Academies Press, 2009). Association of American Medical Colleges, Task Force on Financial Conflicts of Interest in Clinical Research. Protecting Subjects, Preserving Trust, Promoting Progress, *Guidelines for Dealing with Faculty Conflicts of Commitment and Conflicts of Interest in Research* (Washington, DC: AAMC, 2001). Association of American Universities, *Report on Individual and Institutional Financial Conflict of Interest.* Task Force on Research Accountability (Washington, DC: AAU, 2001).

50. J. E. Bekelman, Y. Li, and C. P. Gross, Scope and impact of financial conflicts of interest in biomedical research—A systematic review, *Journal of the American Medical Association* 289, no. 4 (2003): 454–65.

51. E. G. Campbell et al., Institutional academic-industry relationships, *Journal of the American Medical Association* 298, no. 15 (2007): 1779–86.

52. J. Neuman et al., Prevalence of financial conflicts of interest among panel members producing clinical practice guidelines in Canada and United States: cross sectional study, *British Medical Journal* 343 (2011): d5621.

53. A. G. Dunn et al., conflict of interest disclosure in biomedical research: a review of current practices, biases, and the role of public registries in improving transparency, *Research Integrity and Peer Review* 1: 1 (2016). https://researchintegrityjournal.biomedcentral.com/articles/10.1186/s41073-016-0006-7. K. Rasmussen et al., Under-reporting of conflicts of interest among trialists: a cross-sectional study, *Journal of the Royal Society of Medicine* 108, no. 3 (2015): 101–107. M. Roseman et al., Reporting of conflicts of interest from drug trials in Cochrane reviews: cross sectional study, *British Medical Journal* 345 (2012): e5155. R. H. Birkhahn et al., Self-reported financial conflicts of interest during scientific presentations in emergency medicine, *Academic Emergency Medicine* 18, no. 9 (2011): 977–80.

54. Dunn et al., Conflict of interest. J. Lexchin, Sponsorship bias in clinical research, *International Journal of Risk and Safety in Medicine* 24, no. 4 (2012): 233–42.

55. N. Nkansah et al., Randomized trials assessing calcium supplementation in healthy children: relationship between industry sponsorship and study outcomes, *Public Health Nutrition* 12, no. 10 (2009): 1931–37. H. Riaz et al., Impact of funding source on clinical trial results including cardiovascular outcome trials, *American Journal of Cardiology* 116, no. 12 (2015): 1944–47. M. E. Flacco et al., Head-to-head randomized trials are mostly industry sponsored and almost always favor the industry sponsor, *Journal of Clinical Epidemiology* 68, no. 7 (2015): 811–20. D. Mandrioli, C. E. Kearns, and L. A. Bero, Relationship between research outcomes and risk of bias, study sponsorship, and author financial conflicts of interest in reviews of the effects of artificially sweetened beverages on weight outcomes: a systematic review of reviews, *PLOS One* 11, no. 9 (2016): e0162198. L. Bero, Industry sponsorship and research outcome: a Cochrane review, *JAMA Internal Medicine* 173, no. 7 (2013): 580–81. S. Sismondo, Pharmaceutical company funding and its consequences: a qualitative systematic review, *Contemporary Clinical Trials* 29, no. 2 (2008): 109–13.

56. Bero, Industry sponsorship.

57. D. N. Lathyris et al., Industry sponsorship and selection of comparators in randomized clinical trials, *European Journal of Clinical Investigation* 40, no. 2 (2010): 172–82.

58. S. Mathieu et al., Comparison of registered and published primary outcomes in randomized controlled trials, *Journal of the American Medical Association* 302, no. 9 (2009): 977–84.

59. A. G. Dunn et al., Financial conflicts of interest and conclusions about neuraminidase inhibitors for influenza: an analysis of systematic reviews, *Annals of Internal Medicine* 161, no. 7 (2014): 513–18. M. Bes-Rastrollo et al., Financial conflicts of interest and reporting bias regarding the association between sugar-sweetened beverages and weight gain: a systematic review of systematic reviews, *PLOS Medicine* 10, no. 12 (2013): e1001578.

60. J. N. George, S. K. Vesely, and S. H. Woolf, Conflicts of interest and clinical recommendations: comparison of two concurrent clinical practice guidelines for primary immune thrombocytopenia developed by different methods, *American Journal of Medical Quality* 29, no. 1 (2014): 53–60. L. Cosgrove et al., Conflicts of interest and the quality of recommendations in clinical guidelines, *Journal of Evaluation in Clinical Practice* 19, no. 4 (2013): 674–81.

61. Hendriks et al., Trust in science.

62. Besley et al., Perceived conflict of interest. Critchley and Nicol, Understanding the impact.

63. A. Licurse et al., The impact of disclosing financial ties in research and clinical care: a systematic review, *Archives of Internal Medicine* 170, no. 8 (2010): 675–82.

64. R. Proctor, *Golden Holocaust: Origins of the Cigarette Catastrophe and the Case for Abolition* (Berkeley: University of California Press, 2011).

65. R. Horton, Vioxx, the implosion of Merck, and aftershocks at the FDA, *Lancet* 364, no. 9450 (2004): 1995–96.

66. R. A. Pielke, Major change is needed if the IPCC hopes to survive, *Yale Environment 360* (February 25, 2010). http://e360.yale.edu/feature/major_change_is_needed_if_the_ipcc_hopes_to_survive/2244/.

67. Pielke, Major change is needed.

68. K. Brysse et al., Climate change prediction: erring on the side of least drama? *Global Environmental Change-Human and Policy Dimensions* 23, no. 1 (2013): 327–37.

69. World Association of Medical Editors, Conflict of interest. Institute of Medicine, *Conflict of Interest*. Association of American Medical Colleges, *Guidelines*. Association of American Universities, *Report on Individual and Institutional*.

70. I. de Melo-Martín and K. Intemann, How do disclosure policies fail? Let us count the ways, *Federation of American Societies for Experimental Biology Journal* 23, no. 6 (2009): 1638–42. K. C. Elliott, Financial conflicts of interest and criteria for research credibility, *Erkenntnis* 79 (2014): 917–37.

71. Institute of Medicine, *Conflict of Interest*. E. A. Boyd and L. A. Bero, Defining financial conflicts and managing research relationships: an analysis of university conflict of interest committee decisions, *Science and Engineering Ethics* 13, no. 4 (2007)): 415–35.

72. Institute of Medicine, *Conflict of Interest*.

73. Institute of Medicine, *Conflict of Interest*. Association of American Medical Colleges, *Guidelines*. Association of American Universities, *Report on Individual and Institutional*.

74. D. B. Resnik et al., Institutional conflict of interest policies at U.S. academic research institutions, *Academic Medicine* 91, no. 2 (2016): 242–46.

75. A. Nichols-Casebolt and F. L. Macrina, Current perspectives regarding institutional conflict of interest: commentary on "Institutional Conflicts of Interest in Academic Research," *Science and Engineering Ethics* (2015). https://doi.org/10.1007/s11948-015-9703-8. Resnik et al., Institutional conflict.

76. Resnik et al., Institutional conflict.

77. S. A. Gallo, M. Lemaster, and S. R. Glisson, Frequency and type of conflicts of interest in the peer review of basic biomedical research funding applications: self-reporting versus manual detection, *Science and Engineering Ethics* 22, no. 1 (2016): 189–97. de Melo-Martín and Intemann, How do disclosure policies fail? K. D. Elliott, Scientific judgment and the limits of conflict-of-interest policies, *Accountability in Research* 15, no. 1 (2008): 1–29.

78. Brown, The community of science.

79. F. Godlee et al., Journal policy on research funded by the tobacco industry, *Thorax* 68 (2013): 1090–91.

80. B. Zycher et al., Private sector contributions to pharmaceutical science: thirty-five summary case histories, *American Journal of Therapeutics* 17, no. 1 (2010): 101–20. D. Smith and B. MacFadyen, Industry relationships between physicians and professional medical associations: corrupt or essential? *Surgical Endoscopy* 24, no. 2 (2010): 251–53. M. Carrier, Science in the grip of the economy, in M. Carrier et al., eds., *The Challenge of the Social and the Pressure of Practice: Science and Values Revisited* (Pittsburgh, PA: University of Pittsburgh Press, 2008), 217–34.

81. J. Biddle, Lessons from the Vioxx debacle: what the privatization of science can teach us about social epistemology, *Social Epistemology* 21 (2007): 21–39.

82. J. Reiss, In favour of a Millian proposal to reform biomedical research, *Synthese* 177, no. 3 (2010): 427–47.

83. D. A. Zarin, T. Tse, and J. Sheehan, The proposed rule for U.S. clinical trial registration and results submission, *New England Journal of Medicine* 372, no. 2 (2015): 174–80.

84. Reducing our irreproducibility [editorial], *Nature* 496, no. 7446 (2013): 398. Data-access practices strengthened [editorial], *Nature* 515, no. 7527 (2014): 312.

85. I. Simera et al., Transparent and accurate reporting increases reliability, utility, and impact of your research: reporting guidelines and the EQUATOR Network, *BMC Medicine* 8, no. 24 (2010). https://doi.org/10.1186/1741-7015-8-24.

86. Begley and Ioannidis, Reproducibility in science.

87. G. Vogel, Psychologist accused of fraud on "astonishing scale," *Science* 334, no. 6056 (2011): 579.

88. M. McCarthy, Former Duke University oncologist is guilty of research misconduct, U.S. officials find, *British Medical Journal* 351 (2015): h6058.

89. C. Dyer, Duke University settles lawsuits alleging that patients were harmed in chemotherapy trials, *British Medical Journal* 350 (2015): h2559.

90. S. L. George, Research misconduct and data fraud in clinical trials: prevalence and causal factors, *International Journal of Clinical Oncology* 21, no. 1 (2016): 15–21. D. Fanelli, How many scientists fabricate and falsify research? A systematic review and meta-analysis of survey data, *PLOS One* 4, no. 5 (2009): e5738. R. De Vries, M. S. Anderson, and B. C. Martinson, Normal misbehavior: scientists talk about the ethics of research, *Journal of Empirical Research on Human Research Ethics* 1, no. 1 (2006): 43–50. B. C. Martinson, M. S.

Anderson, and R. de Vries, Scientists behaving badly, *Nature* 435, no. 7043 (2005): 737–38. V. Pupovac and D. Fanelli, Scientists admitting to plagiarism: a meta-analysis of surveys, *Science and Engineering Ethics* 21, no. 5 (2015): 1331–52.

D. L. Roberts and F. A. V. St. John, Estimating the prevalence of researcher misconduct: a study of UK academics within biological sciences, *PeerJ* 2: e562 (2014). J. Ana et al., Research misconduct in low- and middle-income countries, *PLOS Medicine* 10, no. 3 (2013): e1001315.

91. S. Godecharle, B. Nemery, and K. Dierickx, Heterogeneity in European research integrity guidance: relying on values or norms? *Journal of Empirical Research on Human Research Ethics* 9, no. 3 (2014): 79–90. George, Research misconduct.

92. F. C. Fang, R. G. Steen, and A. Casadevall, Misconduct accounts for the majority of retracted scientific publications, *Proceedings of the National Academy of Sciences USA* 109, no. 42 (2012): 17028–33.

93. M. L. Grieneisen and M. H. Zhang, A comprehensive survey of retracted articles from the scholarly literature, *PLOS One* 7, no. 10 (2012): e44118

94. R. G. Steen, A. Casadevall, and F. C. Fang, Why has the number of scientific retractions increased? *PLOS One* 8, no. 7 (2013): e68397.

95. Fanelli, How many scientists. Pupovac and Fanelli, Scientists admitting to plagiarism.

96. Fanelli, How many scientists.

97. Pupovac and Fanelli, Scientists admitting to plagiarism.

98. D. Cyranoski, South Korean scandal rocks stem cell community, *Nature Medicine* 12, no. 1 (2006): 4. L. Smith and J. F. Byers, Gene therapy in the post-Gelsinger era, *JONA'S Healthcare Law, Ethics and Regulation* 4, no. 4 (2002): 104–10.

99. A. M. Michalek et al., The costs and underappreciated consequences of research misconduct: a case study, *PLOS Medicine* 7, no. 8 (2010): e1000318.

100. Dyer, Duke University settles lawsuits.

101. Fang et al., Misconduct accounts. Steen et al., Why has the number.

102. D. Fanelli, Why growing retractions are (mostly) a good sign, *PLOS Medicine* 10, no. 12) (2013): e1001563.

103. U. Andersson, Does media coverage of research misconduct impact on public trust in science? A study of news reporting and confidence in research in Sweden 2002–2013, *Observatorio (OBS*)* 9, no. 4 (2015): 15–30.

104. Godecharle et al., Heterogeneity. George, Research misconduct.

105. R. Collier, Is withholding clinical trial results "research misconduct"? *Canadian Medical Association Journal* 187, no. 10 (2015): 724. S. Krimsky, When conflict of interest is a factor in scientific misconduct, *Medicine and Law* 26, no. 3 (2007): 447–63. D. Fanelli, Redefine misconduct as distorted reporting, *Nature* 494, no. 7436 (2013): 149.

106. Fanelli, How many scientists.

107. Roberts and St. John, Estimating the prevalence.

108. Roberts and St. John, Estimating the prevalence.

109. L. Loikith and R. Bauchwitz, The essential need for research misconduct allegation audits, *Science and Engineering Ethics* 22, no. 4 (2016): 1027–49.

110. S. Godecharle, B. Nemery, and K. Dierickx, Guidance on research integrity: no union in Europe, *Lancet* 381, no. 9872 (2013): 1097–98. D. B. Resnik, L. M. Rasmussen, and G. E. Kissling, An international study of research misconduct policies, *Accountability in Research—Policies and Quality Assurance* 22, no. 5 (2015): 249–66.

111. P. R. Patnaik, Scientific misconduct in India: causes and perpetuation, *Science and Engineering Ethics* 22, no. 4 (2016): 1245–49.

112. D. B. Resnik, D. Patrone, and S. Peddada, Research misconduct policies of social science journals and impact factor, *Accountability in Research—Policies and Quality Assurance* 17, no. 2 (2010): 79–84.

113. K. Outterson, Punishing health care fraud: is the GSK settlement sufficient? *New England Journal of Medicine* 367, no. 12 (2012): 1082–85. M. Hvistendahl, Corruption and research fraud send big chill through big pharma in China, *Science* 341, no. 6145 (2013): 445–46. N. Oreskes and E. M. Conway, *Merchants of Doubt: How a Handful of Scientists Obscured the Truth on Issues from Tobacco Smoke to Global Warming* (New York: Bloomsbury, 2010). R. Proctor, *Golden Holocaust: Origins of the Cigarette Catastrophe and the Case for Abolition* (Berkeley: University of California Press, 2011). L. Cosgrove et al., Under the influence: the interplay among industry, publishing, and drug regulation, *Accounts of Chemical Research* 23, no. 5 (2016): 257–79. D. K. Flaherty, Ghost- and guest-authored pharmaceutical industry-sponsored studies: abuse of academic integrity, the peer review system, and public trust, *Annals of Pharmacotherapy* 47, no. 7–8 (2013): 1081–83. Horton, Vioxx, the implosion of Merck.

114. National Academies of Sciences, Engineering, and Medicine, *Fostering Integrity in Research* (Washington, DC: National Academies Press, 2017).

115. D. Fanelli, R. Costas, and V. Lariviere, Misconduct policies, academic culture and career stage, not gender or pressures to publish, affect scientific integrity, *PLOS One* 10, no. 6 (2015): 0127556.

116. C. Seife, Is drug research trustworthy? *Scientific American* 307, no. 6 (2012): 56–63. Loikith and Bauchwitz, The essential need. C. White, Ranjit Chandra: how reputation bamboozled the scientific community, *British Medical Journal* 351 (2015): h5683.

117. S. L. Titus, Evaluating U.S. medical schools' efforts to educate faculty researchers on research integrity and research misconduct policies and procedures, *Accountability in Research—Policies and Quality Assurance* 21, no. 1 (2014): 9–25.

118. R. Smith, Statutory regulation needed to expose and stop medical fraud, *British Medical Journal* 352 (2016): i293.

119. S. Reardon, Uneven response to scientific fraud, *Nature* 523, no. 7559 (2015): 138–9.

120. Outterson, Punishing health care fraud.

121. P. C. Gotzsche, Big pharma often commits corporate crime, and this must be stopped, *British Medical Journal* 345 (2012): e8462.

122. Fanelli et al., Misconduct policies. M. S. Davis, M. Riske-Morris, and S. R. Diaz, Causal factors implicated in research misconduct: evidence from ORI case files, *Science and Engineering Ethics* 13, no. 4 (2007): 395–414. J. K. Tijdink et al., How do scientists perceive the current publication culture? A qualitative focus group interview study among Dutch biomedical researchers, *BMJ Open* 6, no. 2 (2016): e008681. Martinson et al., Scientists behaving badly. A. Casadevall and F. C. Fang, Reforming science: methodological and cultural reforms, *Infection and Immunity* 80, no. 3 (2012): 891–96.

123. Tijdink et al., How do scientists.

124. J. M. DuBois et al., Lessons from researcher rehab, *Nature* 534, no. 7606 (2016): 173–75.

125. M. S. Anderson et al., The perverse effects of competition on scientists' work and relationships, *Science and Engineering Ethics* 13, no. 4 (2007): 437–61.

126. Anderson et al., The perverse effects. F. C. Fang and A. Casadevall, Competitive science: is competition ruining science? *Infection and Immunity* 83, no. 4 (2015): 1229–33.

127. National Academies of Sciences, Engineering, and Medicine, *Fostering Integrity in Research* (Washington, DC: National Academies Press, 2017). Fanelli et al., Misconduct policies.

128. Resnik et al., An international study.

129. National Academies of Sciences, Engineering, and Medicine, *Fostering Integrity in Research*. D. S. Kornfeld and S. L. Titus, Stop ignoring misconduct, *Nature* 537, no. 7618 (2016): 29–30.

130. D. E. Wright, S. L. Titus, and J. B. Cornelison, Mentoring and research misconduct: an analysis of research mentoring in closed ORI cases, *Science and Engineering Ethics* 14, no. 3 (2008): 323–36.

131. National Academies of Sciences, Engineering, and Medicine, *Fostering Integrity in Research*. Kornfeld and Titus, Stop ignoring misconduct.

Chapter 9

1. Q. Schiermeier, Storm clouds gather over leaked climate e-mails, *Nature* 462, no. 7272 (2009): 397.

2. L. Hickman and J Randerson, Climate sceptics claim leaked emails are evidence of collusion among scientists, *The Guardian* (2009). https://www.theguardian.com/environment/2009/nov/20/climate-sceptics-hackers-leaked-emails.

3. M. Russell et al., The independent climate change emails review (2010, p. 11). http://www.cce-review.org/pdf/final%20report.pdf.

4. Russell et al., The independent climate change emails review, 14.

5. A. Leiserowitz et al., Climategate, public opinion, and the loss of trust, *American Behavioral Scientist* 57, no. 6 (2013): 818–37. E. Maibach et al., The legacy of climategate: undermining or revitalizing climate science and policy? *Wiley Interdisciplinary Reviews—Climate Change* 3, no. 3 (2012): 289–95.

6. G.-E. Séralini et al., Long-term toxicity of a Roundup herbicide and a Roundup-tolerant genetically modified maize [retracted], *Food and Chemical Toxicology* 50, no. 11) (2012): 4221–31.

7. D. Butler, Hyped GM maize study faces growing scrutiny, *Nature* 490, no. 7419 (2012): 158.

8. Butler, Hyped GM maize study.

9. Poison postures [editorial], *Nature* 489, no. 7417 (2012): 474.

10. G. Arjo et al., Plurality of opinion, scientific discourse and pseudoscience: an in depth analysis of the Séralini et al. study claiming that Roundup Ready corn or the herbicide Roundup causes cancer in rats, *Transgenic Research* 22, no. 2 (2013): 255–67. L. Ollivier, A comment on Séralini, G.-E. et al., "Long term toxicity of a Roundup herbicide and a Roundup-tolerant genetically modified maize," *Food and Chemical Toxicology* 53 (2013): 458. R. Pilu, Comment on "Long term toxicity of a Roundup herbicide and a Roundup-tolerant genetically modified maize" by Séralini et al., *Food and Chemical Toxicology* 53 (2013): 454. D. Sanders et al., Comment on "Long term toxicity of a Roundup herbicide and a Roundup-tolerant genetically modified maize" by Séralini et al., *Food and Chemical Toxicology* 53 (2013): 450–53. M. Tester, Comment

on "Long term toxicity of a Roundup herbicide and a Roundup-tolerant genetically modified maize" by Séralini et al., *Food and Chemical Toxicology* 53 (2013): 457. A. Trewavas, Comment on "Long term toxicity of a Roundup herbicide and a Roundup-tolerant genetically modified maize" by Séralini et al., *Food and Chemical Toxicology* 53 (2013): 449.

11. H. Miller and B. Chassy, Scientists smell a rat in fraudulent genetic engineering study, *Forbes*, September 12, 2012. https://geneticliteracyproject.org/2012/09/27/scientists-smell-a-rat-in-fraudulent-genetic-engineering-study/.

12. Data-access practices strengthened [editorial], *Nature* 515, no. 7527 (2014): 312. The decision to withdraw the article because of the inconclusiveness of the results has also been criticized. See D. B. Resnik, Retracting inconclusive research: lessons from the Séralini GM maize feeding study, *Journal of Agricultural Environmental Ethics* 28, no. 4 (2015): 621–33.

13. G.-E. Séralini et al., Republished study: long-term toxicity of a Roundup herbicide and a Roundup-tolerant genetically modified maize, *Environmental Sciences Europe* 26, no. 1 (2014). https://doi.org/10.1186/s12302-014-0014-5.

14. Gilles-Eric Séralini, Robin Mesnage, Nicolas Defarge, and Joël Spiroux de Vendômois, Conflicts of interests, confidentiality and censorship in health risk assessment: the example of an herbicide and a GMO, *Environmental Sciences Europe* 26, no. 13 (2014). https://doi.org/10.1186/s12302-014-0013-6.

15. Pew Research Center, Public and scientists' views on science and society (2015). www.pewinternet.org/files/2015/01/PI_ScienceandSociety_Report_012915.pdf.

16. Pew Research Center, The politics of climate (2016). www.pewinternet.org/2016/10/04/the-politics-of-climate/.

17. Pew Research Center, The politics of climate.

18. M. Brown, Values in science beyond underdetermination and inductive risk, *Philosophy of Science* 80, no. 5 (2014): 829–39. S. Haack, *Manifesto of a Passionate Moderate: Unfashionable Essays* (Chicago: University of Chicago Press, 1998). G. Betz, In defense of the value free ideal, *European Journal of the Philosophy of Science* 3, no. 2 (2013): 207–20. H. Douglas, *Science, Policy, and the Value-Free Ideal* (Pittsburgh, PA: University of Pittsburgh Press, 2009). E. Anderson, Uses of value judgments in science: a general argument, with lessons from a case study of feminist research on divorce, *Hypatia* 19, no. 1 (2004): 1–24. K. C. Elliott, *A Tapestry of Values: An Introduction to Values in Science* (New York: Oxford University Press, 2017).

19. N. Roll-Hansen, *The Lysenko Effect: The Politics of Science* (Amherst, NY: Humanity Books, 2005). L. R. Graham, *Science and Philosophy in the Soviet Union* (New York: Knopf, 1972).

20. Elliott, *A Tapestry of Values*.

21. M. D. Gordin, How Lysenkoism became pseudoscience: Dobzhansky to Velikovsky, *Journal of the History of Biology* 45, no. 3 (2012): 443–68.

22. A. Fausto-Sterling, *Myths of Gender: Biological Theories About Women and Men* (New York: Basic Books, 1992). S. B. Hrdy, Empathy, polyandry and the myth of the coy female, in R. Bleier, ed., *Feminist Approaches to Science* (New York: Pergamon, 1986), 119–46. E. Martin, The egg and the sperm: how science has constructed a romance based on stereotypical male-female relationships, *Signs* 16, no. 3 (1991): 485–501. A. Wylie and L. H. Nelson, Coming to terms with the values of science: insights from feminist science studies

scholarship, in H. Kincaid, J. Dupre, and A. Wylie, eds., *Value-Free Science? Ideals and Illusions* (Oxford: Oxford University Press, 2007), 58–86. S. S. Richardson, *Sex Itself: The Search for Male and Female in the Human Genome* (Chicago: University of Chicago Press, 2013). H. E. Longino, *Studying Human Behavior: How Scientists Investigate Aggression and Sexuality* (Chicago: University of Chicago Press, 2013).

23. Martin, *The egg and the sperm*.

24. S. J. Gould, *The Mismeasure of Man* (New York: Norton, 1981).

25. R. A. Kerr, "Arctic Armageddon" needs more science, less hype, *Science* 329, no. 5992 (2010): 620–21.

26. A. Patt and S Dessai, Communicating uncertainty: lessons learned and suggestions for climate change assessment, *Comptes Rendus Geoscience* 337, no. 4 (2005): 425–41.

27. P. J. Michaels, How to manufacture a climate consensus, *Wall Street Journal*, December 17, 2009. www.wsj.com/articles/SB10001424052748704398304574598230426037244. R. A. Pielke, When scientists politicize science: making sense of controversy over the Skeptical Environmentalist, *Environmental Science and Policy* 7, no. 5 (2004): 405–17.

28. N. Oreskes and E. M. Conway, *Merchants of Doubt: How a Handful of Scientists Obscured the Truth on Issues from Tobacco Smoke to Global Warming* (New York: Bloomsbury, 2010). D. Michaels, *Doubt Is Their Product: How Industry's Assault on Science Threatens Your Health* (Oxford and New York: Oxford University Press, 2008). J. Biddle and A. Leuschner, Climate skepticism and the manufacture of doubt: can dissent in science be epistemically detrimental? *European Journal for Philosophy of Science* 5, no. 3 (2015): 261–78. P. A. Offit, *Deadly Choices: How the Anti-Vaccine Movement Threatens Us All* (New York: Basic Books, 2011).

29. Pew Research Center, *The politics of climate*.

30. M. Dorato, Epistemic and nonepistemic values in science, in P. Machamer and G. Wolters, eds., *Science, Values, and Objectivity* (Pittsburgh, PA: University of Pittsburgh Press, 2004), 52–77. Betz, In defense of the value free ideal. R. Hudson, Why we should not reject the value-free ideal of science, *Perspectives on Science* 24, no. 2 (2016): 167–91.

31. S. Mitchell, The prescribed and proscribed values in science policy, in P. Machamer and G. Wolters, eds., *Science, Values and Objectivity* (Pittsburgh, PA: University of Pittsburgh Press, 2004), 245–55. Dorato, Epistemic and nonepistemic values. R. Lackey, Science, scientists, and policy advocacy, *Conservation Biology* 21, no. 1 (2007): 12–17.

32. R. A. Pielke, *The Honest Broker: Making Sense of Science in Policy and Politics* (Cambridge, UK and New York: Cambridge University Press, 2007).

33. B. Miller, Science, values, and pragmatic encroachment on knowledge, *European Journal for Philosophy of Science* 4, no. 2 (2014): 253–70. H. E. Longino, *Science as Social Knowledge: Values and Objectivity in Scientific Inquiry* (Princeton, NJ: Princeton University Press, 1990). M. Solomon, *Social Empiricism* (Cambridge, MA: MIT Press, 2001). Douglas, *Science, Policy, and the Value-Free Ideal*. J. Dupré, Fact and value, in H. Kincaid et al., eds., *Value-Free Science? Ideals and Illusions* (Oxford: Oxford University Press, 2007), 27–41.

34. Miller, Science, values, and pragmatic encroachment. J. Biddle, State of the field: transient underdetermination and values in science, *Studies in History and Philosophy of Science* 44, no. 1 (2013): 124–33.

35. I. de Melo-Martín and K. Intemann, The risk of using inductive risk to challenge the value-free ideal, *Philosophy of Science* 83, no. 4 (2016): 500–20. D. J. Hicks, A new direction for science and values, *Synthese* 191, no. 14 (2014): 3271–95. Brown, Values in science.

K. Elliott, Douglas on values: from indirect roles to multiple goals, *Studies in History and Philosophy of Science* 44, no. 3 (2013): 375–83. K. Intemann and I. de Melo-Martín, Social values and scientific evidence: the case of the HPV vaccines, *Biology and Philosophy* 25, no. 2 (2010): 203–13. Douglas, *Science, Policy, and the Value-Free Ideal.*

36. Anderson, Uses of value judgments in science. J. Callicott et al., Current normative concepts in conservation, *Conservation Biology* 13, no. 1 (1999): 22–35. Dupré, Fact and value. K. Elliott, The ethical significance of language in the environmental sciences: case studies from pollution research, *Ethics, Place, & Environment* 12, no. 2 (2009): 157–73. D. Ludwig, Ontological choices and the value-free ideal, *Erkenntnis* 81, no. 6 (2016): 1253–72. K. Shrader-Frechette, *Risk and Rationality: Philosophical Foundations for Populist Reforms* (Berkeley: University of California Press, 1991).

37. Anderson, Uses of value judgments in science. Shrader-Frechette, *Risk and Rationality.*

38. Intemann and de Melo-Martín, Social values and scientific evidence.

39. Ludwig 2016.

40. F. Wickson and B. Wynne, Ethics of science for policy in the environmental governance of biotechnology: MON819 maize in Europe, *Ethics, Policy and Environment* 15, no. 3 (2012): 321–40.

41. Wickson and Wynne, Ethics of science.

42. Intemann and de Melo-Martín, Social values and scientific evidence. Wickson and Wynne, Ethics of science.

43. H. Longino, Gender, politics, and the theoretical virtues, *Synthese* 104, no. 3 (1995): 383–97. Wylie and Nelson, Coming to terms with the values of science. K. Intemann, Distinguishing between legitimate and illegitimate values in climate modeling, *European Journal of Philosophy of Science* 5, no. 2 (2015): 217–32.

44. Longino, Gender, politics.

45. W. Parker, Confirmation and adequacy for purpose in climate modelling, *Aristotelian Society Supplementary Volume* 83, no. 1 (2009): 233–49. T. Mauritsen et al., Tuning the climate of a global model, *Journal of Advances in Modeling Earth Systems* 4 (2012): 1–18.

46. Mauritsen et al., Tuning the climate.

47. Intemann, Distinguishing between legitimate and illegitimate values.

48. Shrader-Frechette, *Risk and Rationality.* H. Douglas, Inductive risk and values in science, *Philosophy of Science* 67, no. 4 (2000): 559–79. Douglas, *Science, Policy, and the Value-Free Ideal.*

49. L. O. Mearns, Quantification of uncertainties of future climate change: challenges and applications, *Philosophy of Science* 77, no. 5 (2010): 998–1011.

50. J. Biddle and E. Winsberg, Value judgements and the estimation of uncertainty in climate modeling, in P. D. Magnus and J. Busch, eds., *New Waves in Philosophy of Science* (Basingstoke, UK: Palgrave MacMillan, 2010), 172–97.

51. Biddle and Winsberg, Value judgements. E. Winsberg, Values and uncertainties in the predictions of global climate models, *Kennedy Institute of Ethics Journal* 22, no. 2 (2012): 111–37.

52. F. S. Collins and L. A. Tabak, NIH plans to enhance reproducibility, *Nature* 505, no. 7485 (2014): 612–13. Reducing our irreproducibility [editorial], *Nature* 496, no. 7446 (2013): 398.

53. Where are the data? [editorial], *Nature* 537, no. 7619 (2016): 138.

54. H. E. Longino, *The Fate of Knowledge* (Princeton, NJ: Princeton University Press, 2002). Longino, *Science as Social Knowledge*. M. Solomon, Norms of epistemic diversity, *Episteme: A Journal of Social Epistemology* 3, no. 1 (2006): 23–36. K. Borgerson, Amending and defending critical contextual empiricism, *European Journal for Philosophy of Science* 1, no. 3 (2011): 435–49. M. Carrier, Values and objectivity in science: value-ladenness, pluralism and the epistemic attitude, *Science and Education* 22, no. 10 (2013): 2547–68.

55. Solomon, Norms of epistemic diversity.

56. Carrier, Values and objectivity. H. Lacey, *Is Science Value Free? Values and Scientific Understanding* (London and New York: Routledge, 1999). N. Rescher, *Pluralism: Against the Demand for Consensus* (New York: Oxford University Press, 1993).

57. G. Gauchat, Politicization of science in the public sphere: a study of public trust in the United States, 1974 to 2010, *American Sociological Review* 77, no. 2 (2012): 167–87. A. M. McCright et al., Political ideology and views about climate change in the European Union, *Environmental Politics* 25, no. 2 (2016): 338–58. L. C. Hamilton, J. Hartter, and K. Saito, Trust in scientists on climate change and vaccines, *Sage Open* 5, no. 3 (2015). doi: 2158244015602752.

58. J. K. Amory, Male contraception, *Fertility and Sterility* 106, no. 6 (2016): 1303–309.

59. S. Brown, "They think it's all up to the girls": gender, risk and responsibility for contraception, *Culture Health and Sexuality* 17, no. 3 (2015): 312–25.

60. A. Agarwal A southern perspective on curbing global climate change, in S. H. Schneider et al., eds., *Climate Change Policy: A Survey* (Washington, DC: Island Press, 2002), 375–91. S. H. Schneider, Integrated assessment of global climate change: transparent rational tool for policymaking or opaque screen hiding value-laden assumptions? *Environmental Modeling and Assessment* 2 (1997): 229–49.

61. W. N. Adger et al., Are there social limits to adaptation to climate change? *Climatic Change* 93, no. 3–4 (2009): 335–54.

62. K. Shrader-Frechette, Hydrogeology and framing questions having policy consequences, *Philosophy of Science* 64, no. 4 (1997): S149–60.

63. T. Wilholt, Epistemic trust in science, *British Journal for the Philosophy of Science* 64, no. 2 (2013): 233–53.

64. Pew Research Center, The politics of climate. McCright et al., Political ideology and views about climate change. Hamilton et al., Trust in scientists. Gauchat, Politicization of science.

65. B. R. McFadden Examining the gap between science and public opinion about genetically modified food and global warming, *PLOS One* 11, no. 11 (2016): e0166140. M. A. Ranney and D. Clark, Climate change conceptual change: Scientific information can transform attitudes, *Topics in Cognitive Science* 8, no. 1 (2016): 49–75. R. E. Dunlap and A. M. McCright, A widening gap—Republican and Democratic views on climate change, *Environment* 50, no. 5 (2008): 26–35. K. S. Fielding and M. J. Hornsey, A social identity analysis of climate change and environmental attitudes and behaviors: insights and opportunities, *Frontiers in Psychology* 7, no. 121 (2016): doi: 10.3389/fpsyg.2016.00121. A. Zia and A. M. Todd, Evaluating the effects of ideology on public understanding of climate change science: how to improve communication across ideological divides? *Public*

Understanding of Science 19, no. 6 (2010): 743–61. D. M. Kahan, Ideology, motivated reasoning, and cognitive reflection, *Judgment and Decision Making* 8, no. 4 (2013): 407–24.

66. Pielke, *The Honest Broker*. Betz, In defense of the value free ideal.

67. Douglas, *Science, Policy, and the Value-Free Ideal*. K. Elliott and D. McKaughan, Nonepistemic values and the multiple goals of science, *Philosophy of Science* 81, no. 1 (2014): 1–21. Elliott, Douglas on values.

68. Elliott, Douglas on values.

69. de Melo-Martín and Intemann, The risk of using inductive risk

70. de Melo-Martín and Intemann, The risk of using inductive risk.

71. K. Shrader-Frechette, *Taking Action, Saving Lives: Our Duties to Protect Environmental and Public Health* (Oxford and New York: Oxford University Press, 2007). J. L. Shirk et al., Public participation in scientific research: a framework for deliberate design, *Ecology and Society* 17, no. 2 (2012): 29. http://dx.doi.org/10.5751/ES-04705-170229. R. Bonney et al., Can citizen science enhance public understanding of science? *Public Understanding of Science* 25, no. 1 (2016): 2–16.

72. A. K. Adams et al., Using community advisory boards to reduce environmental barriers to health in American Indian Communities, Wisconsin, 2007–2012, *Preventing Chronic Disease* 11 (September 2014): 140014. R. M. Gonzalez-Guarda et al., Advancing nursing science through community advisory boards: working effectively across diverse communities, *Advances in Nursing Science* 40, no. 3 (2016): 278–88. B. Pratt et al., Exploitation and community engagement: can community advisory boards successfully assume a role minimising exploitation in international research? *Developing World Bioethics* 15, no. 1 (2015): 18–26.

73. R. M. Pinto, A. Y. Spector, and P. A. Valera, Exploring group dynamics for integrating scientific and experiential knowledge in community advisory boards for HIV research, *Aids Care–Psychological and Socio-Medical Aspects of Aids/HIV* 23, no. 8 (2011): 1006–13. S. F. Morin et al., Community consultation in HIV prevention research: a study of community advisory boards at 6 research sites, *Journal of Acquired Immune Deficiency Syndromes* 33, no. 4 (2003): 513–20. M. R. Isler et al., Across the miles: process and impacts of collaboration with a rural community advisory board in HIV research, *Progress in Community Health Partnerships—Research Education and Action* 9, no. 1 (2015): 41–48.

74. P. Kloprogge and J. P. van der Sluijs, The inclusion of stakeholder knowledge and perspectives in integrated assessment of climate change, *Climatic Change* 75, no. 3 (2006): 359–89. S. Tang and S. Dessai, Usable science? The U.K. Climate projections 2009 and decision support for adaptation planning, *Weather, Climate, and Society* 4 (2012): 300–13. C. J. Kirchhoff et al., Actionable knowledge for environmental decision making: broadening the usability of climate science, *Annual Review of Environment and Resources* 38 (2013): 393–414.

75. Tang and Dessai, Usable science?

76. S. G. Harding, *Sciences from Below: Feminisms, Postcolonialities, and Modernities*. Next Wave: New Directions in Women's Studies (Durham, NC: Duke University Press, 2008). K. Intemann, Why diversity matters: understanding and applying the diversity component of the National Science Foundation's broader impacts criterion, *Social Epistemology* 23, no. 3–4 (2009): 249–66.

77. Douglas, *Science, Policy, and the Value-Free Ideal*.

Chapter 10

1. M. Kuntz, The postmodern assault on science: if all truths are equal, who cares what science has to say? *EMBO Reports* 13, no. 10 (2012): 885–89.

2. P. Hackett and D. Carroll, Regulatory hurdles for agriculture GMOs, *Science* 347, no. 6228 (2015): 1324.

3. G. Harris, Plan to widen availability of morning-after pill is rejected, *New York Times*, December 7, 2011. http://www.nytimes.com/2011/12/08/health/policy/sebelius-overrules-fda-on-freer-sale-of-emergency-contraceptives.html.

4. Politics and the morning-after pill [editorial], *New York Times*, December 7, 2011. http://www.nytimes.com/2011/12/08/opinion/politics-and-the-morning-after-pill.html.

5. P. Belluck, Judge strikes down age limits on morning-after pill, *New York Times*, April 5, 2013. http://www.nytimes.com/2013/04/06/health/judge-orders-fda-to-make-morning-after-pill-available-over-the-counter-for-all-ages.html.

6. M. A. Schreurs, The Paris Climate Agreement and the three largest emitters: China, the United States, and the European Union, *Politics and Governance* 4, no. 3 (2016): 219–23.

7. Schreurs, The Paris Climate Agreement.

8. S. Pruitt and L. Strange 2016, The climate-change gang, *National Review*. May 17, 2016.

9. I. de Melo-Martín and K. Intemann, Scientific dissent and public policy. Is targeting dissent a reasonable way to protect sound policy decisions? *EMBO Reports* 14, no. 3 (2013): 231–35. K. Intemann and I. de Melo-Martín. Regulating scientific research: should scientists be left alone? *Federation of American Societies for Experimental Biology Journal* 22, no. 3 (2008): 654–58. D. J. Hicks, Scientific controversies as proxy politics, *Issues in Science and Technology* 33, no. 2 (2017): 67–72. R. A. Pielke, *The Honest Broker: Making Sense of Science in Policy and Politics* (Cambridge, UK, and New York: Cambridge University Press, 2007).

10. N. Oreskes and E. M. Conway, *Merchants of Doubt: How a Handful of Scientists Obscured the Truth on Issues from Tobacco Smoke to Global Warming* (New York: Bloomsbury, 2010). T. O. McGarity and W. Wagner, *Bending Science: How Special Interests Corrupt Public Health Research* (Cambridge, MA: Harvard University Press, 2008). D. Harker, *Creating Scientific Controversies: Uncertainty and Bias in Science and Society* (Cambridge, UK: Cambridge University Press, 2015). M. Kuntz, The postmodern assault on science: if all truths are equal, who cares what science has to say? *EMBO Reports* 13, no. 10 (2012): 885–89.

11. Note that we are not denying that NID can indeed produce confusion or lead people to have false beliefs about the state of the evidence or the degree of consensus in some scientific areas with controversial policy consequences. Our claim is that such need not be the case and that one can appropriately disagree with certain policy options supported by the best scientific evidence available without rejecting the scientific evidence in question.

12. S. Drews and J. C. J. M. van den Bergh, What explains public support for climate policies? A review of empirical and experimental studies, *Climate Policy* 16, no. 7 (2016): 855–76. N. Harring and S. C. Jagers, Should we trust in values? Explaining public support for pro-environmental taxes, *Sustainability* 5, no. 1 (2013): 210–27. A. Nilsson and A. Biel, Acceptance of climate change policy measures: role framing and value guidance, *European Environment* 18, no. 4 (2008): 203–15. N. Smith and A. Leiserowitz, The role of emotion in global warming policy support and opposition, *Risk Analysis* 34, no. 5

(2014): 937–48. A. Leiserowitz, Climate change risk perception and policy preferences: the role of affect, imagery, and values, *Climatic Change* 77, no. 1–2 (2006): 45–72.

13. R. Shwom et al., Understanding U.S. public support for domestic climate change policies, *Global Environmental Change—Human and Policy Dimensions* 20, no. 3 (2010): 472–82. R. E. O'Connor et al., Who wants to reduce greenhouse gas emissions? *Social Science Quarterly* 83, no. 1 (2002): 1–17.

14. D. Sarewitz, How science makes environmental controversies worse, *Environmental Science & Policy* 7, no. 5 (2004): 385–403.

15. Although through the chapter we make a distinction between scientific evidence and value judgments, we do not mean to imply that scientific evidence is value free or that value judgments involve no empirical evidence. As we discussed in chapter 9, nonepistemic or contextual values are part and parcel of science. We are calling attention here to the fact that scientific evidence, even when it is appropriately value laden, is not sufficient to determine public policies. A variety of additional value judgments are also necessary to guide policy decisions.

16. S. H. Fritts and L. N. Carbyn, Population viability, nature- reserves, and the outlook for gray wolf conservation in north-America, *Restoration Ecology* 3, no. 1 (1995): 26–38.

17. M. A. Wilson, The wolf in Yellowstone: science, symbol, or politics? Deconstructing the conflict between environmentalism and wise use, *Society & Natural Resources* 10, no. 5 (1997): 453–68.

18. D. A. McNaught, Wolves in Yellowstone: park visitors respond, *Wildlife Society Bulletin* 15, no. 4 (1987): 518–21. A. J. Bath, The public and wolf reintroduction in Yellowstone National Park, *Society and Natural Resources* 2, no. 4 (1989): 297–306.

19. B. N. Beisher, Are ranchers legitimately trying to save their hides or are they just crying wolf? What issues must be resolved before wolf reintroduction to Yellowstone National Park proceeds, *Land and Water Law Review* 29, no. 2 (1994): 417–65.

20. S. H. Fritts, E. E. Bangs, and J. F. Gore, The relationship of wolf recovery to habitat conservation and biodiversity in the northwestern United States, *Landscape and Urban Planning* 28, no. 1 (1994): 23–32.

21. L. D. Mech, The challenge and opportunity of recovering wolf populations, *Conservation Biology* 9, no. 2 (1995): 270–78.

22. J. T. Bruskotter et al., Are gray wolves endangered in the Northern Rocky Mountains? A role for social science in listing determinations, *Bioscience* 60, no. 11 (2010): 941–48. B. J. Bergstrom et al., The Northern Rocky Mountain gray wolf is not yet recovered, *Bioscience* 59, no. 11 (2009): 991–99.

23. Sarewitz, How science makes environmental controversies worse.

24. N. Oreskes, D. A. Stainforth, and L. A. Smith, Adaptation to global warming: do climate models tell us what we need to know? *Philosophy of Science* 77, no. 5 (2010): 1012–28.

25. J. Abrams et al., Value orientation and forest management: the forest health debate, *Environmental Management* 36, no. 4 (2005): 495–505.

26. Abrams et al., Value orientation.

27. Abrams et al., Value orientation. D. B. Tindall, Social values and the contingent nature of public opinion and attitudes about forests, *Forestry Chronicle* 79, no. 3 (2003): 692–705.

28. B. S. Steel, P. List, and B. Shindler, Conflicting values about federal forests: a comparison of national and Oregon publics, *Society and Natural Resources* 7, no. 2 (1994): 137–53. T. E. Kolb, M. R. Wagner, and W. W. Covington, Concepts of forest health: utilitarian and ecosystem perspectives, *Journal of Forestry* 92, no. 7 (1994): 10–15.

29. Kolb et al., Concepts of forest health.

30. R. Pielke et al., Lifting the taboo on adaptation, *Nature* 445, no. 7128 (2007): 597–98. M. Parry, J. Lowe, and C. Hanson, Overshoot, adapt and recover, *Nature* 458, no. 7242 (2009): 1102–103. Oreskes et al., Adaptation to global warming.

31. P. M. Kelly and W. N. Adger, Theory and practice in assessing vulnerability to climate change and facilitating adaptation, *Climatic Change* 47, no. 4 (2000): 325–52.

32. W. N. Adger et al., This must be the place: underrepresentation of identity and meaning in climate change decision-making, *Global Environmental Politics* 11, no. 2 (2011): 1–25.

33. Adger et al., This must be the place.

34. H. McLachlan and D. Forster, The safety of home birth: is the evidence good enough? *Canadian Medical Association Journal* 181, no. 6–7 (2009): 359–60.

35. American Medical Association (AMA), Resolution 205: Home Deliveries (2008). http://www.ama-assn.org/ama1/pub/upload/mm/471/205.doc.

36. McLachlan and Forster, The safety of home birth. D. Young, Home birth in the United States: action and reaction, *Birth* 35, no. 4 (2008): 263–65.

37. AMA, Resolution 205. A. Grunebaum et al., Early and total neonatal mortality in relation to birth setting in the United States, 2006–2009, *American Journal of Obstetrics and Gynecology* 211, no. 4 (2014): 390-e. N. K. Lowe, The "authorities" resolve against home birth, *Journal of Obstetrical Gynecology—Neonatal Nursing* 38, no. 1 (2009): 1–3.

38. I. de Melo-Martín and K. Intemann, Interpreting evidence: why values can matter as much as science, *Perspectives in Biology and Medicine* 55, no. 1 (2012): 59–70.

39. E. Tracy, Does home birth empower women or imperil them and their babies? *OBG Management* 21, no. 8 (2009): 44–52. AMA, Resolution 205.

40. Tracy, Does home birth empower women.

41. J. S. Cohain, What would an evidence-based statement on homebirths from ACOG say? *Midwifery Today/International Midwife* 86 (2008): 32–33.

42. M.-R. Jouhki, Choosing home birth: the women's perspective, *Women and Birth* 25, no. 4 (2012): E56–E61.

43. P. A. Offit, *Deadly Choices: How the Anti-Vaccine Movement Threatens Us All* (New York: Basic Books, 2011).

44. I. A. Harmsen et al., Why parents refuse childhood vaccination: a qualitative study using online focus groups, *BMC Public Health* 13, no. 1 (2013): 1183. M. J. Goldenberg, Public misunderstanding of science? Reframing the problem of vaccine hesitancy, *Perspectives on Science* 24, no. 5 (2016): 552–81.

45. M. A. Largent, *Vaccine: The Debate in Modern America* (Baltimore, MD: Johns Hopkins University Press, 2012).

46. Gallup, U.S. concern about global warming at eight-year high—poll (2016). www.gallup.com/poll/190010/concern-global-warming-eight-year-high.aspx.

47. Gallup, U.S. concern. A. Leiserowitz et al., *Climate Change in the American Mind: March 2016* (New Haven, CT: Yale University Press, 2016). P. D. Howe et al.,

Geographic variation in opinions on climate change at state and local scales in the USA, *Nature Climate Change* 5, no. 6 (2015): 596–603.

48. J. Jacquet et al., Intra- and intergenerational discounting in the climate game, *Nature Climate Change* 3, no. 12 (2013): 1025–28.

49. L. Scruggs and S. Benegal, Declining public concern about climate change: can we blame the great recession? *Global Environmental Change—Human and Policy Dimensions* 22, no. 2 (2012): 505–15.

50. A. J. Hoffman, Talking past each other? Cultural framing of skeptical and con- vinced logics in the climate change debate, *Organization and Environment* 24, no. 1 (2011): 3–33. Y. Heath and R. Gifford, Free-market ideology and environmental degra- dation: the case of belief in global climate change, *Environment and Behavior* 38, no. 1 (2006): 48–71.

51. UN Food and Agriculture Organization, *The State of Food Insecurity in the World 2013* (Rome: UN Food and Agriculture Organization, 2013).

52. J. Berman et al., Can the world afford to ignore biotechnology solutions that address food insecurity? *Plant Molecular Biology* 83, no. 1–2 (2013): 5–19. S. S. Gill et al., Genetic engineering of crops: a ray of hope for enhanced food security, *Plant Signaling & Behavior* 9, no. 3 (2014): e28545.

53. Gill et al., Genetic engineering of crops. R. Paarlberg, GMO foods and crops: Africa's choice, *New Biotechnology* 27, no. 5 (2010): 609–13.

54. G.-E. Séralini et al., Republished study: long-term toxicity of a Roundup herbi- cide and a Roundup-tolerant genetically modified maize, *Environmental Sciences Europe* 26, no. 1 (2014). https://doi.org/10.1186/s12302-014-0014-5. M. Cuhra, Review of GMO safety assessment studies: glyphosate residues in Roundup Ready crops is an ignored issue, *Environmental Sciences Europe* 27 (2015): 20.

55. H. Lacey, *Values and Objectivity in Science* (Lanham, MD: Rowman & Littlefield, 2005). I. de Melo-Martín and Z. Meghani, Beyond risk: a more realistic risk-benefit analysis of agricultural biotechnologies, *EMBO Reports* 9, no. 4 (2008): 302–306. D. J. Hicks, Scientific controversies as proxy politics, *Issues in Science and Technology* 33, no. 2 (2017): 67–72.

56. National Academies of Sciences Engineering, and Medicine (NASEM), *Genetically Engineered Crops: Experiences and Prospects*. U.S. Committee on Genetically Engineered Crops: Past Experience and Future Prospects. (Washington, DC: National Academies Press, 2016).

57. H. Azadi et al., Genetically modified crops and small-scale farmers: main opportunities and challenges, *Critical Reviews in Biotechnology* 36, no. 3 (2016): 434–46. K. Gerasimova, Debates on genetically modified crops in the context of sustainable develop- ment, *Science and Engineering Ethics* 22, no. 2 (2016): 525–47. V. Shiva, Limiting corporate power and cultivating interdependence: a strategic plan for the environment, *Tikkun* 30, no. 2 (2015): 26–27. Lacey, *Values and Objectivity in Science*.

58. A. D. M. Briggs et al., Assessing the impact on chronic disease of incorporating the societal cost of greenhouse gases into the price of food: an econometric and comparative risk assessment modelling study, *BMJ Open* 3, no. 10 (2013): e003543.

59. F. N. Tubiello et al., The FAOSTAT database of greenhouse gas emissions from ag- riculture, *Environmental Research Letters* 8, no. 1 (2013): 015009.

Chapter 11

1. Intergovernmental Panel on Climate Change (IPCC), *Climate Change 2013: The Physical Science Basis*. Working Group I contribution to the fifth assessment report (Geneva: Intergovernmental Panel on Climate Change, 2013).

2. J. Cook et al., Quantifying the consensus on anthropogenic global warming in the scientific literature, *Environmental Research Letters* 8, no. 2 (2013): 024024. J. Cook et al., Consensus on consensus: a synthesis of consensus estimates on human-caused global warming, *Environmental Research Letters* 11, no. 4 (2016): 048002.

3. S. Pruitt and L. Strange, The climate-change gang, *National Review*, May 17, 2016.

4. C. Davenport, Climate change and the incoming Trump government, *New York Times*, December 19, 2016.

5. R. Perry, *Fed Up! Our Fight to Save America from Washington* (New York: Little, Brown, 2010), 189.

6. B. Tranter and K. Booth, Scepticism in a changing climate: a cross-national study, *Global Environmental Change—Human and Policy Dimensions* 33 (2015): 154–64. M. Darby, UK public ignorant of climate science consensus—poll, *Climate Change News*, August 27, 2014. http://www.climatechangenews.com/2014/08/27/uk-public-ignorant-of-climate-science-consensus-poll/.

7. R. Proctor, *Golden Holocaust: Origins of the Cigarette Catastrophe and the Case for Abolition* (Berkeley: University of California Press, 2011). D. Michaels, *Doubt Is Their Product: How Industry's Assault on Science Threatens Your Health* (Oxford and New York: Oxford University Press, 2008). T. O. McGarity and W. Wagner, *Bending Science: How Special Interests Corrupt Public Health Research* (Cambridge, MA: Harvard University Press, 2008).

8. N. Oreskes and E. M. Conway, *Merchants of Doubt: How a Handful of Scientists Obscured the Truth on Issues from Tobacco Smoke to Global Warming* (New York: Bloomsbury, 2010). J. Biddle and A. Leuschner, Climate skepticism and the manufacture of doubt: can dissent in science be epistemically detrimental? *European Journal for Philosophy of Science* 5, no. 3 (2015): 261–78. D. Harker, *Creating Scientific Controversies: Uncertainty and Bias in Science and Society* (Cambridge, UK: Cambridge University Press, 2015). S. Lewandowsky et al., The role of conspiracist ideation and worldviews in predicting rejection of science, *PLOS One* 8, no. 10 (2013): e75637. M. Aklin and J. Urpelainen, Perceptions of scientific dissent undermine public support for environmental policy, *Environmental Science and Policy* 38 (2014): 173–77.

9. Michaels, *Doubt Is Their Product*. Oreskes and Conway, *Merchants of Doubt*. Harker, *Creating Scientific Controversies*. Biddle and Leuschner, Climate skepticism and the manufacture of doubt. A. Leuschner. Is It Appropriate to 'Target' Inappropriate Dissent? On the Normative Consequences of Climate Skepticism. *Synthese* 195, no. 3 (2018): 1255–1271.

10. S. L. Van der Linden et al., The scientific consensus on climate change as a gateway belief: experimental evidence, *PLOS One* 10, no. 2 (2015): e0118489. Lewandowsky et al., The role of conspiracist ideation. Cook et al., Quantifying the consensus on anthropogenic global warming. Cook et al., Consensus on consensus. S. L. van der Linden et al., Highlighting consensus among medical scientists increases public support for vaccines: evidence from a randomized experiment, *BMC Public Health* 15 (2015): 1207.

11. K. S. Shrader-Frechette, *Tainted: How Philosophy of Science Can Expose Bad Science* (Oxford and New York: Oxford University Press, 2014).

12. Michaels, *Doubt Is Their Product*. Oreskes and Conway, *Merchants of Doubt*. M. E. Mann, *The Hockey Stick and the Climate Wars: Dispatches from the Front Lines* (New York: Columbia University Press, 2012). Shrader-Frechette, *Tainted*.

13. Michaels, *Doubt Is Their Product*. Oreskes and Conway, *Merchants of Doubt*. Proctor, *Golden Holocaust*. R. Brulle, Institutionalizing delay: foundation funding and the creation of U.S. climate change counter-movement organizations, *Climatic Change* 122, no. 4 (2014): 681–94.

14. A reminder here is in order. Recall that the concern about the reliability of criteria for NID is tied to the aims to which they will be put to use. Insofar as such aims include implementing strategies that could limit or silence dissenting views, we think it is reasonable to require that criteria for NID be highly reliable so as to avoid the possibility of silencing epistemically sound dissent. If the use of such criteria is unlikely to limit epistemically valuable dissent, then a concern with their reliability is less pressing.

15. Oreskes and Conway, *Merchants of Doubt*. Biddle and Leuschner, Climate skepticism and the manufacture of doubt. P. Kitcher, *Science in a Democratic Society* (Amherst, NY: Prometheus, 2011). H. E. Longino, *The Fate of Knowledge* (Princeton, NJ: Princeton University Press, 2002). D. K. Flaherty, The vaccine-autism connection: a public health crisis caused by unethical medical practices and fraudulent science, *Annals of Pharmacotherapy* 45, no. 10 (2011): 1302–304. J. S. Gerber and P. A. Offit, Vaccines and autism: a tale of shifting hypotheses, *Clinical Infectious Diseases* 48, no. 4 (2009): 456–61.

16. Kitcher, *Science in a Democratic Society*. Harker, *Creating Scientific Controversies*.

17. A. Hilbeck et al., No scientific consensus on GMO safety, *Environmental Sciences Europe* 27, no. 4 (2015): https://doi.org/10.1186/s12302-014-0034-1. S. Krimsky. An Illusory Consensus behind GMO Health Assessment. *Science Technology & Human Values*, 40, no. 6 (2015): 883–914.

18. A. Agarwal, A southern perspective on curbing global climate change, in S. H. Schneider and A. Rosencranz and J. O. Niles, eds., *Climate Change Policy: A Survey* (Washington, DC: Island Press, 2002), 375–91. M. Hulme and S. Dessai, Negotiating future climates for public policy: a critical assessment of the development of climate scenarios for the UK, *Environmental Science and Policy* 11, no. 1 (2008): 54–70.

19. J. Biddle et al., Epistemic corruption and manufactured doubt: the case of climate science, *Public Affairs Quarterly* 31, no. 3 (2017): 165–88.

20. Longino, *The Fate of Knowledge*. H. E. Longino, *Science as Social Knowledge: Values and Objectivity in Scientific Inquiry* (Princeton, NJ: Princeton University Press, 1990).

21. D. M. Kahan, Ideology, motivated reasoning, and cognitive reflection, *Judgment and Decision Making* 8, no. 4 (2013): 407–24. D. M. Kahan, Climate-science communication and the measurement problem, *Political Psychology* 36 (2015): 1–43. B. R. McFadden, Examining the gap between science and public opinion about genetically modified food and global warming, *PLOS One* 11, no. 11 (2016): e0166140. M. A. Ranney and D. Clark, Climate change conceptual change: scientific information can transform attitudes, *Topics in Cognitive Science* 8, no. 1 (2016): 49–75. K. S. Fielding and M. J. Hornsey, A social identity analysis of climate change and environmental attitudes and behaviors: insights and opportunities, *Frontiers in Psychology* 7, no. 121 (2016): doi: 10.3389/fpsyg.2016.00121.

A. Zia and A. M. Todd, Evaluating the effects of ideology on public understanding of climate change science: how to improve communication across ideological divides? *Public Understanding of Science* 19, no. 6 (2010): 743–61. E. Dubé et al., Vaccine hesitancy, vaccine refusal and the anti-vaccine movement: influence, impact and implications, *Expert Review of Vaccines* 14, no. 1 (2015): 99–117. M. Navin, *Values and Vaccine Refusal: Hard Questions in Ethics, Epistemology, and Health Care* (New York: Routledge, 2016).

22. S. Drews and J. C. J. M. Van den Bergh, What explains public support for climate policies? A review of empirical and experimental studies, *Climate Policy* 16, no. 7 (2016): 855–76. N. Harring and S. C. Jagers, Should we trust in values? Explaining public support for pro-environmental taxes, *Sustainability* 5, no. 1 (2013): 210–27. A. Nilsson and A. Biel, Acceptance of climate change policy measures: role framing and value guidance, *European Environment* 18, no. 4 (2008): 203–15.

23. N. Smith and A. Leiserowitz, The role of emotion in global warming policy support and opposition, *Risk Analysis* 34, no. 5 (2014): 937–48. A. Leiserowitz, Climate change risk perception and policy preferences: the role of affect, imagery, and values, *Climatic Change* 77, no. 1–2 (2006): 45–72.

24. R. Shwom et al., Understanding U.S. public support for domestic climate change policies, *Global Environmental Change–Human and Policy Dimensions* 20, no. 3 (2010): 472–82. R. E. O'Connor et al., Who wants to reduce greenhouse gas emissions? *Social Science Quarterly* 83, no. 1 (2002): 1–17.

25. B. Cai, T. A. Cameron, and G. R. Gerdes, Distributional preferences and the incidence of costs and benefits in climate change policy, *Environmental and Resource Economics* 46, no. 4 (2010): 429–58. J. J. Lee and T. A. Cameron, Popular support for climate change mitigation: evidence from a general population mail survey, *Environmental & Resource Economics* 41, no. 2 (2008): 223–48.

26. Cai et al., Distributional preferences. Lee and Cameron, Popular support.

27. E. Parasidis, Public health law and institutional vaccine skepticism, *Journal of Health and Political Policy Law* 41, no. 6 (2016): 1137–49.

28. R. Grundmann, The legacy of climategate: revitalizing or undermining climate science and policy? *Wiley Interdisciplinary Reviews—Climate Change* 3, no. 3 (2012): 281–88. M. Ryghaug and T. M. Skjolsvold, The global warming of climate science: climategate and the construction of scientific facts, *International Studies in the Philosophy of Science* 24, no. 3 (2010): 287–307.

29. J. Beatty, Consensus: sometimes it doesn't add up, in S. Gissis, E. Lamm, and A. Shavit, eds., *Landscapes of Collectivity in the Life Sciences* (Cambridge, MA: MIT Press, 2018).

30. Beatty, Consensus.

31. P. Kloprogge and J. P. van der Sluijs, The inclusion of stakeholder knowledge and perspectives in integrated assessment of climate change, *Climatic Change* 75, no. 3 (2006): 359–89. S. Tang and S. Dessai, Usable science? The U.K. Climate projections 2009 and decision support for adaptation planning, *Weather, Climate, and Society* 4 (2012): 300–13. C. J. Kirchhoff et al., Actionable knowledge for environmental decision making: broadening the usability of climate science, *Annual Review of Environment and Resources* 38 (2013): 393–414.

INDEX

Printed in the USA/Agawam, MA
July 8, 2019

706661.001